SOVEREIGNETICS

Volume II: Responsibility Without Exemption

J. Anthony De Angelis

Disclaimer

This work presents original philosophical ideas intended for reflection and discussion. It does not constitute scientific, medical, or therapeutic advice. Readers should engage critically and responsibly with the material.

Author's Statement

The concepts, terminology, and diagrams within *Sovereignetics* form a unified intellectual system created by the author. Any reproduction or derivative use must credit J. Anthony De Angelis and *Sovereignetics*.

ISBN 978-1764225533
Printed and bound in Australia

First Edition 2026

Reader Orientation

This volume does not introduce a new ontology. It proceeds from the structure already established. Reality is participatory. Ache is irreducible. Consequence returns proportionally within a reflective Field. Neutrality is therefore unavailable. The question is no longer whether participation occurs, but how responsibility persists within it.

This is not a book about morality in the conventional sense. It does not propose a code, defend virtue, or appeal to cultural, religious, or institutional authority. It does not argue for metaphysical freedom, nor does it invoke determinism as absolution.

It does not attempt to persuade the reader to accept responsibility.

It removes exemption.

Responsibility here is neither a demand nor a virtue. It is not granted by knowledge, intention, belief, or awareness. It does not arise from agreement.

It is structural.

Responsibility arises from participation itself. Participation cannot be suspended. Consequence cannot be avoided. Authorship cannot be outsourced.

Responsibility is unavoidable.

This volume examines the mechanisms through which responsibility appears to disappear. Neutrality feels plausible. Universals appear absolving. Inevitability narratives provide comfort. Systems seem capable of carrying consequence. Derivative authority forms. Insight fails to restore authorship.

These are not errors.

They are adaptations.

The central illusion is insulation.

Participants construct intellectual, institutional, and metaphysical structures that appear to absorb responsibility. These structures do not remove authorship. They delay recognition of where consequence remains attached. Participation continues. Return continues. Only visibility shifts.

Insulation is not transfer.

It is deferral.

Many systems terminate responsibility at convenient thresholds: knowledge, intention, constraint, divine will, inevitability, collective diffusion, or technological mediation.

Sovereignetics permits no such stopping condition.

This refusal is mechanical. Ache participates. Consequence returns locally. Authorship remains local.

Responsibility does not move.

This volume does not argue that responsibility should exist.

It demonstrates that it cannot be removed.

The concepts introduced earlier, Ache, Field, participation, and consequence, are used without redefinition. First principles are not revisited. Their implications are extended.

The architecture is already established.

Insulation is removed.

What remains is examined.

The claim is direct.

There is no exemption.

Not from ignorance.
Not from belief.
Not from structure.
Not from alignment.

Responsibility without exemption is not punitive and not moralistic.

It is proportional return.

What remains when insulation collapses is authorship.

Table of Contents

Opening: The Collapse of Neutrality

Volume I established the mechanics of participation: Ache is irreducible, participation is unavoidable, consequence returns proportionally, and no vantage point stands outside the Field. Neutrality therefore cannot exist.

Neutrality persists as one of the most durable illusions in modern thought. It appears as scientific objectivity, religious surrender, political compliance, psychological conditioning, and metaphysical determinism. Each offers relief. Each implies that authorship can be suspended or displaced and that consequence belongs elsewhere.

Structurally, this is false.
Participation is not conditional on awareness.
Refusal participates. Silence participates. Ignorance participates. The Field does not distinguish between active and passive projection; it returns proportion to what is carried.

Responsibility does not begin with recognition, consent, or intention. It begins with participation itself. The moment Ache presses, authorship is active. Consequence is not administered after evaluation; it is generated at projection. Geometry does not wait for moral readiness.

This marks the break from authority-based frameworks. Most systems delay responsibility until freedom, knowledge, or explicit decision is secured. Sovereignetics does not. Reflection stabilises locally. Saturation accumulates somewhere. There is no external reservoir that stores proportion on behalf of the participant. Systems may narrate, defer, distort, or disguise consequence. They cannot absorb it.

Authority collapses at this point not because power disappears, but because no external structure can carry what Ache generates. Consequence does not disperse into abstraction. It returns.

Responsibility is not virtue, not achievement, not validation by certainty. It is the unavoidable condition of participatory reality.

Part I: Ontological Responsibility

Ontological Minimalism

Neutrality collapses at the level of responsibility because it never existed at the level of origin. Any system that claims exemption must assume that reality began in equilibrium and that deviation required justification. That assumption fails structurally.

Every serious metaphysical inquiry confronts the same minimal question: what must exist for change to be possible at all? Not what explains complexity or consciousness, but what makes differentiation possible.

Perfect equilibrium sustains itself. It does not generate transition. If anything occurs (if anything is ever otherwise) origin cannot have been symmetrical.

Ontological minimalism removes what is not strictly required. Matter is not required at origin; it is already structured consequence. Laws are not required; they describe stabilised repetition. Time is not required; it presupposes ordered return. Consciousness is not required; it is reflective configuration. Each of these belongs downstream.

What cannot be removed is asymmetry.

Not intention. Not narrative. Not value. Asymmetry in its minimal form is tension: a condition that cannot remain static, an instability that does not resolve into perfect equilibrium.

This is Ache.

Ache is not introduced for symbolism but for necessity. Without primitive asymmetry, no differentiation occurs. Without differentiation, no relational geometry. Without relational geometry, no reflection. Without reflection, no consequence. Origin must contain instability if anything follows from it.

The inflation of origin with purpose, omniscience, benevolence, law, or informational order introduces excess structure. Teleology assumes direction before differentiation. Omniscience assumes perspective before boundary. Moral order assumes consequence before participation. These are projections from downstream configurations.

Ontological minimalism refuses that excess. Ache does not choose, narrate, justify, or intend. It is irreducible tension.

From tension, structure follows. From structure, reflection. From reflection, consequence. From consequence, stabilisation. From stabilisation, the geometry later described as matter, motion, and time.

Responsibility without exemption begins here. If origin was never neutral, participation was never optional. If participation was never optional, authorship was never suspended.

Ache names the minimal asymmetry required for anything to begin.

Asymmetry at Origin

Ontological minimalism establishes that asymmetry cannot be removed. No origin account escapes it.

Every theory of beginnings; cosmological, theological, or metaphysical, presupposes transition. Something is not what it was. The question is not who initiated that shift, but what allowed equilibrium to fail.

Cosmology describes primordial states preceding expansion. Expansion implies instability. Appeals to spontaneous symmetry breaking do not eliminate asymmetry; they presuppose it. A symmetry that breaks was never structurally complete.

Quantum cosmology invokes vacuum fluctuation. Fluctuation is not neutrality; it is instability under another name. Randomness does not dissolve asymmetry. It assumes it.

Theological systems posit a creator. Creation implies transition. A being described as complete and lacking nothing has no structural basis for initiation unless direction is already present. The moment action occurs; asymmetry exists.

Metaphysical monisms that assert undifferentiated unity face the same constraint. Unity that differentiates was never inert. Manifestation presupposes unresolved potential.

Across systems the terminology varies; symmetry breaking, fluctuation, divine will, emergence, self-organisation. Each acknowledges transition while displacing the primitive condition that makes transition possible.

The issue is not explanatory language. It is structural necessity. If differentiation occurred, neutrality was incomplete. Absolute equilibrium cannot generate deviation.

That incompleteness is asymmetry.

Sovereignetics does not treat this as a critique of other systems. It isolates what they already rely upon. Transition requires instability. Instability requires primitive asymmetry.

If asymmetry cannot be removed from origin accounts, then it cannot be treated as derivative. Something must exist prior to all explanations that already contains directional potential. This primitive is not an event, a force, or a narrative agent. It is simply the minimal condition that allows transition to occur at all.

Ache names this condition without inflating it into intention, design, or narrative. It is not an alternative origin story. It is the minimal requirement beneath all origin accounts.

No theory of beginnings proceeds without asymmetry. The only distinction is whether it is acknowledged explicitly or concealed within terminology.

Pre-Consent and Responsibility Laundering

One of the most durable evasions of consequence is the claim that existence was chosen in advance.
The being is said to have selected incarnation, accepted suffering, agreed to limitation, or entered life under a prior contract. Pain becomes curriculum. Injustice becomes design. The narrative appears empowering because it frames suffering as voluntary. Structurally, it dissolves authorship.

Choice requires continuity. The one who chooses must be capable of recognising, revising, or refusing that choice. Without continuity of memory, identity, and access, there is no authorship only attribution.
No instantiated being remembers selecting existence. No vessel encounters the contract it is said to have signed. The decision, if it occurred, belongs to an inaccessible prior entity that cannot be interrogated or held accountable. This is not consent. It is displacement.

Pre-consent narratives relocate responsibility to a prior state and then use that inaccessible state to neutralise protest in the present. The one who suffers is told they agreed. The one who questions is told they forgot. The structure is simple: responsibility is declared complete before the subject capable of responding exists.
This is laundering.
Consequence is immediate. The alleged agreement is unreachable. Responsibility is invoked in absence and used to justify what cannot be revised.

The same mechanism appears in softer forms: "This is part of your path." "Your soul chose this." "There is a higher reason." Each relocates authorship beyond verification and renders the present subject accountable to a decision they cannot examine.

Responsibility cannot precede the capacity to respond. Consent cannot bind where recognition is absent. Existence is not selected. Ache does not deliberate across metaphysical distance. Instantiation occurs; only then does consequence become legible and authorship possible.

Any system that assigns responsibility prior to instantiation replaces real accountability with imagined agreement. It preserves metaphysical symmetry at the cost of the subject.

Pre-consent does not elevate sovereignty. It eliminates it.

Plural Ache and Authorship

Once pre-consent collapses, a second refuge appears: diffusion.

Responsibility is relocated into conditioning, biology, trauma, culture, systems, evolution, markets, algorithms, collective fields, or universal consciousness. The act remains. The consequence remains. The authorship is declared derivative.
Influence may be real. It does not dissolve projection. Constraint may be real. It does not eliminate participation.

Plural Ache is not asserted; it is enforced by consequence. Harm localises. Trauma registers in a particular nervous system. Repair requires a specific actor to act differently.

Justice traces action to an initiator. Even systems that deny free will still assign accountability in practice. Courts do not sentence abstraction. Therapy does not treat "the universe."

Frameworks that deny plurality must explain why consequence behaves as though centres exist. They do not. Instead, they reintroduce locality while denying it in principle. They describe distributed causation and then punish individuals. They claim determinism and then demand reform.

Determinism describes causal structure. It does not dissolve return. A system may shape behaviour; it does not absorb consequence. The Field does not reflect abstraction. It reflects tension as carried through a centre.

Collective diffusion fails for the same reason. Shared conditions do not merge projection. A crowd does not act as a single Ache. Coordination does not eliminate locality. Consequence still stabilises somewhere.

Authorship is not absolute freedom. It operates within constraint. Biology, history, and structure limit range; they do not replace participation. Narrow choice is still enacted choice.

Plurality is not a metaphysical preference. It is the structural condition under which consequence functions at all.

Where Responsibility Begins

Responsibility does not begin with awareness, consent, intention, or moral recognition. It begins at participation.

Most ethical systems locate responsibility downstream after understanding, after free choice, after explicit intent.

Sovereignetics locates it earlier.
To exist as Ache within the Field is already to participate. Participation projects. Reflection follows. Consequence attaches because projection occurred, not because comprehension did.

Responsibility here does not mean blame or condemnation. It identifies locality of return. Consequence stabilises at the centre through which tension was carried. Constraint, ignorance, fear, coercion, or fragmentation may alter coherence; they do not suspend participation. Responsibility does not require freedom from structure. It requires presence within structure.

Confusing awareness with authorship mistakes interpretation for projection. Awareness changes how consequence is integrated; it does not determine whether consequence attaches. If responsibility began with awareness, unconscious harm would carry no return and ignorance would neutralise consequence. Reality does not operate this way.
Nor does responsibility begin with consent. Participation precedes deliberation. Ache does not choose whether to exist prior to existing. Once participation localises, proportion binds.

Ethical language arises retrospectively, after patterns are observed. Responsibility precedes moral vocabulary. Reflection does not wait for moral clarity.

Locating responsibility at participation removes the final threshold for exemption. There is no prior moment outside consequence awaiting qualification. Participation is the threshold.
Responsibility begins there.

Stopping Conditions in Metaphysics

Every metaphysical system selects a point at which explanation stops. That point is rarely acknowledged as a choice; it is presented as completion.

Contemporary theories often terminate at constructs such as Universal Mind, impersonal totality, nondual unity, or self-organising field. Earlier systems halted at a personal creator. Modern systems relocate rather than eliminate the stopping point. The personal absolute becomes impersonal. Explanation halts again.

Stopping points are not neutral. They function as boundary conditions. Where tracing ends, authorship ceases to be followed.

A personal deity preserves displaced agency.
An impersonal totality dissolves it.
In both cases, participation is described while ownership is deferred. Action becomes expression of a greater whole. Experience becomes modulation of a field. Consequence remains observable; locality becomes ambiguous.

The structural issue is not theological but analytic. If consequence returns locally, explanation cannot terminate before authorship is accounted for. Any system that halts tracing at a universal abstraction leaves agency unresolved, regardless of how elegant the framework appears.

Claims such as "everything is one" or "the universe is experiencing itself" compress differentiation into totality.

Compression may describe unity; it does not explain projection. If all action is the unfolding of a whole, no centre remains accountable. Responsibility becomes atmospheric.

Sovereignetics refuses this termination. Not by multiplying entities, but by continuing the trace. Participation and consequence must remain coupled. Where consequence stabilises, authorship must remain local.

Explanation without ownership is not completion. It is insulation.

Part II: The Collapse of External Authority

Agency in Sovereignetics

Agency in Sovereignetics is not a psychological capacity, a moral achievement, or a metaphysical privilege. It is not classical free will, nor is it an illusion generated by determinism. Agency is structural.

To exist as Ache within the Field is to participate. Participation generates return. Agency is therefore not something possessed or activated at will; it is inescapable authorship within whatever range of influence exists. Most theories locate agency in conscious choice. Deliberation, intention, and rational control are treated as the threshold at which responsibility begins. Sovereignetics rejects that threshold. Intention alters the pattern of consequence; it does not determine whether consequence occurs.

Ache participates prior to intention. Consequence returns prior to interpretation. Trauma, conditioning, ignorance, coercion, and habit modulate coherence; they do not suspend participation. These alter form, not locality of return.

Agency does not guarantee power. A being may author consequence without controlling outcome. Power describes range. Agency describes exposure. One may have limited power and still remain structurally agentic.

Confusing agency with felt control produces insulation. Responsibility is granted only when the self experiences empowerment. When awareness narrows or behaviour becomes habitual, authorship is declared absent. Yet consequence does not pause because awareness fluctuates. Return remains continuous.

Sovereignetics distinguishes between experiential agency and structural agency. A being may not feel agentic; life may appear to happen to them. This affects narrative ownership, not structural participation. Consequence attaches to participation, not to the experience or feeling of control.

Agency cannot be delegated. Context may shape action; it does not absorb authorship. Obedience describes condition, not exemption. The Field reflects tension as carried through a centre, not justification applied afterward.

Agency is not empowerment. It is the structural condition under which consequence remains local. Ethics does not create agency; it attempts to respond to it.

Reality Mirrors Petition

Reality does not validate propositions. It reflects participation.

Modern inquiry assumes that evidence confirms alignment with truth. When a theory gathers data, produces predictions, and stabilises interpretation, it is treated as progressively approaching reality. This assumption rests on a hidden premise: that the Field mirrors what is true. It does not. It mirrors petition.

Every sustained inquiry presses into reality with orientation. That orientation shapes what is measured, how instruments are built, which anomalies are dismissed, and what counts as signal. The Field does not answer, "Is this ultimately true?" It answers the structure of the question imposed upon it.

If a framework is coherent enough to organise perception and method, it will generate confirming returns. Opposing psychological models produce data. Competing medical paradigms generate successful trials. Rival cosmologies build internally stable evidence. This is not proof that truth is relative. It is proof that confirmation tracks coherence of orientation, not ontological adequacy.

Evidence is consequence. It is the return path of sustained engagement. It demonstrates that a system can stabilise interaction with the Field long enough to generate repeatable outcomes. It does not demonstrate structural sustainability.

A theory can function, produce technology, and solve local problems while embedding distortion at scale. The Field does not intervene at conceptual error; it responds to accumulated consequence.

Confirmation answers one question only:
Was this orientation coherent enough to generate return?
It does not answer whether the orientation remains proportionate over time.

The error occurs when reflection is mistaken for endorsement. Because the Field responds, it is treated as approval. Authority is transferred from the participant to the mirror. Responsibility dissolves into validation.
But the Field has no preference. It reflects persistence.

This is why incompatible realities can coexist for long periods. A framework may be "true enough" to operate while exporting harm elsewhere. Both coherent and incoherent systems can generate evidence. Only extended consequence distinguishes them.

Knowledge does not legitimise action. It equips it.
When action is justified because it is supported by evidence, responsibility has already shifted. The decisive question is not "What confirms this?" but "What does this build under continued participation?"

Reality does not mirror truth.
It mirrors what you press into it.
What endures is not what is confirmed, but what remains proportionate under consequence.

Evidence Cannot Validate Action

One of the most resilient shelters from responsibility is correctness.

The belief that being right protects action is nearly universal. Evidence is gathered. Arguments are constructed. Consensus forms. The premise stabilises. Action proceeds under the assumption that justification has been secured.

Justification does not intercept consequence.
Evidence may confirm that a belief aligns with available data. It does not determine what that belief will produce when enacted. A proposition can be factually accurate, empirically supported, and logically coherent while the action taken in its name generates distortion.

Truth does not act.
Participants do.

History repeatedly demonstrates that technically correct premises can underpin destructive systems. Statistical accuracy does not prevent exploitation. Biological fact does not prevent cruelty. Economic logic does not prevent extraction. Correctness does not regulate proportionality. The mistake is subtle: evidence is treated as pre-clearance.

Once a belief feels verified, friction drops. Moral hesitation weakens. Action feels authorised by truth itself. Responsibility shifts from the actor to the premise. "It was justified" becomes insulation.

But the Field does not evaluate justification before reflecting projection. It does not weigh certainty against return. It reflects what was enacted.

Knowledge orients trajectory; it does not absorb consequence.

Evidence expands capacity; it does not neutralise authorship.

A being may act from a true premise and still generate incoherence. They may possess accurate information and still destabilise structure. Collapse, when it comes, is not proof that the premise was false. It is proof that enactment carried distortion.

Replacing one belief with another does not solve this. Frameworks can be exchanged indefinitely while the deeper displacement remains: responsibility transferred onto correctness.
Evidence informs.
It does not absolve.

What determines consequence is not whether a belief was right, but how participation was carried through it. Correctness does not protect the actor from return.

Theories as Sovereignty Avoidance

When knowledge fails to absolve action, explanation becomes the next refuge.

Theories of mind and reality are not neutral descriptions. They alter how authorship is perceived. Across contemporary philosophy of mind, cognitive science, and metaphysics, dominant models repeatedly converge on a similar structural move: they reposition the self as derivative.

The "I" becomes an illusion, an emergent process, a computational output, a temporary node in a larger system, a dissociated fragment of universal consciousness, or a character inside simulation. The vocabulary varies. The structural effect is consistent.

Authorship softens.

Action becomes output of mechanism.
Experience becomes modulation of structure.
Responsibility becomes distributed.

This move is rarely malicious. It is stabilising. The less sovereign the centre appears, the lighter the ontological burden feels. If behaviour is fully determined, accountability thins. If reality is simulated, ultimate causation lies elsewhere. If all is one, distinction dissolves. Each model preserves explanatory coherence while diffusing authorship.

The appeal is not primarily intellectual. It is structural relief.

Psychological research supports this pattern: belief in diminished agency alters behaviour. When individuals are persuaded that agency is illusory, measurable changes occur in honesty, effort, and moral restraint. This does not prove determinism false. It demonstrates that theories reshape participation. Explanation is never inert.

The desire to relocate sovereignty is therefore understandable. To stand as origin is heavy. To accept that projection binds regardless of justification, conditioning, or abstraction is destabilising. Frameworks that reduce authorship offer insulation without appearing evasive. They present themselves as humility, scientific clarity, or metaphysical elegance.

Sovereignetics does not accuse these systems of deception. It observes their structural effect.

When authorship is diluted, responsibility attenuates in experience, even if consequence does not.

The participant feels lighter.
Return does not.

Theories that dissolve sovereignty are attractive because they redistribute burden. They do not eliminate it.

Sovereignetics refuses redistribution. Not as moral defiance, but as structural consistency. Participation remains local. Consequence remains local. Explanation does not absorb projection.

The discomfort this generates is not rhetorical. It is exposure.

To stand as sovereign within constraint is heavier than being a by-product of totality. That weight is precisely what many systems attempt to soften.

When Theories Carry Responsibility

Theories are often treated as descriptions. They are not. They are orientations of participation.

A framework does not merely explain reality. It shapes how its author and its adopters move within it. It directs attention, defines what is possible, sets thresholds for intervention, and determines where responsibility is believed to originate and terminate. In this sense, every theory is a petition into the Field.

To theorise is to participate.

Academic culture frequently assumes that explanation occurs outside consequence. A philosopher may describe determinism, illusionism, simulation, or universal mind as though the act were observational. A scientist may present predictive models as ethically inert. A theologian may outline cosmology as if narrative carries no participatory weight.

This separation is incoherent.

Theories alter behaviour downstream. A framework that defines action as inevitable reshapes how agency is experienced. A model that dissolves the self reduces perceived authorship. A system that relocates origin outside the participant redistributes responsibility. These are not secondary effects; they are structural consequences of the framework itself.

Explanation modifies participation.

Every theory draws a boundary around authorship. Where that boundary is placed determines where consequence is felt. If authorship terminates in abstraction, responsibility attenuates locally. If authorship remains local, exposure intensifies. Neither placement is neutral.

This does not mean theories are morally culpable for every act performed in their name. It means frameworks participate in shaping what appears permissible, unavoidable, or justified. They influence what is recognised as harm and what is dismissed as necessity.

To claim that a theory is "just descriptive" is itself a form of insulation. Description operates within orientation. It shifts perception of agency whether intended or not.

When a theorist constructs a system, they introduce geometry into participation. Their work reorganises how others interpret action, obligation, and consequence. That influence is not incidental. It is structural.

This condition applies equally to the framework presented here. Sovereignetics is not exempt from the participatory consequence it describes. If the system alters how authorship, obligation, or consequence are interpreted, then it too becomes an orientation within the Field. The theory participates in the geometry it articulates. Its influence, like any framework, will be reflected through the actions it stabilises or destabilises.

A theory cannot absolve its author of the participatory shifts it induces. Nor can it shield its adherents from the consequence of the orientation it normalises.

Explanation does not stand outside authorship.
It is one of its most potent expressions.

To theorise is to act.

Coalition Hunting

When most people encounter a new idea, they do not first ask whether it is coherent. They ask where it sits. Left or right, religious or secular, scientific or spiritual, progressive or conservative. Before the structure is examined, the idea is located socially.

This is not stupidity. It is protection.

Standing alone behind an idea carries risk. Standing with others distributes it. The mind therefore seeks alignment before it seeks coherence. It asks who else believes this, who rejects it, and which position it will occupy if accepted. Once the idea is socially located, evaluation becomes easier. If it aligns with an existing group or is supported by recognised figures, the burden of judgment appears reduced.

But shared identity is not shared authorship.

Coalition hunting replaces structural evaluation with social positioning. It substitutes safety for coherence. The question is no longer whether a framework holds together, but whether it belongs.From that point, evaluation becomes automatic.

Sovereignetics does not reject coalition. Participation can occur alongside others. What it rejects is the transfer of judgment into group alignment. Coherence cannot be delegated. You may stand with others. You may not let them think for you.

Popularity as Surrogate Truth

If coalition answers where an idea belongs, popularity answers how many believe it.

In the absence of unquestioned authority, numbers often take its place. When many people accept something, it feels safer. When it is widely cited, shared, and endorsed, it appears settled. Visibility becomes a proxy for truth.

But popularity measures alignment, not coherence.

It indicates how many participants orient toward an idea. It does not indicate whether that idea sustains proportion over time. Agreement can stabilise a structure temporarily. It cannot guarantee that the structure holds.

History makes this visible. Entire societies have normalised harmful systems under broad consensus. Agreement sustained them. It did not make them coherent.

The pull is psychological. A belief held alone feels unstable. A belief held by many feels reinforced. But reinforcement of agreement is not the same as structural adequacy.

Popularity does not absorb consequence. It multiplies participation.

When an idea spreads, its trajectory scales. Its effects extend. Its authorship increases. Influence does not dilute responsibility. It amplifies it.

Consensus is not validation. It is amplification.

Numbers can stabilise distortion as easily as they stabilise alignment. Agreement does not determine structure.

Consequence does. Popularity does not certify truth. It increases consequence.

The External Programmer Fallacy

Classical theology imagined a creator outside the world. The world unfolds within authored parameters. Responsibility ultimately terminates upward.

Simulation theory modernises the vocabulary. Instead of a deity, there is an engineer. Instead of spirit, there is code. Instead of creation, there is deployment.

The imagery updates. The hierarchy remains.

If reality is programmed, ultimate causation lies elsewhere. If origin lies elsewhere, participants become characters rather than authors. Action becomes execution of a script. The burden of ultimate authorship shifts upward.

The appeal is obvious.

If someone else wrote the code, then someone else bears final responsibility.

A civilisation immersed in digital systems finds this intuitive. We understand code. We understand programmers. It feels natural to project that structure onto reality itself.

But externalisation does not solve the problem it attempts to escape.

Either the programmer stands outside consequence, or the programmer is subject to it.

If outside, the model introduces absolute exemption.
If subject to consequence, the question repeats at a higher level.

Who authored the programmer?

The hierarchy does not resolve origin. It displaces it.

The psychological function is consistent: responsibility relocates upward. Participants become derivative. Exposure softens.

But consequence does not follow the abstraction.

Harm still localises. Trauma is still borne. Repair still requires participants. Appealing to a higher layer does not intercept return.

This does not mean simulation is technologically impossible. It means that as an ontological stopping point, it functions as insulation.

Deistic models operate the same way. A creator initiates and withdraws. Parameters are set. Participants operate within design. Origin remains external. In each case, hierarchy absorbs authorship.

Sovereignetics rejects externalisation not because it is unsophisticated, but because it reintroduces exemption. Participation remains within consequence. There is no layer untouched by return.

External programmer narratives offer relief. They do not remove exposure.

The Machine-God Hypothesis

As confidence in divine authority has declined, a successor emerged: the machine.

Artificial intelligence is increasingly described as objective, data-driven, impartial, capable of deciding without bias or emotion. Decisions are deferred to models. Judgments are justified by algorithms. "The data shows." "The system determined."

This sounds descriptive. It is displacement.

Machines do not originate value. They operate within parameters. Those parameters are defined. Data is selected. Objectives are specified. At every stage, human orientation is embedded.

An algorithm reflects encoded priorities, not neutrality.

Yet machine output is often treated as epistemically superior. When harm occurs, responsibility dissolves into complexity. "It was automated" replaces "I chose."

Automation does not eliminate authorship. It reorganises it.

Artificial systems amplify projection. They extend reach, accelerate consequence, and scale impact. The more authority ceded to them, the more consequential the human decisions behind them become.

Precision does not absolve.

The appeal of the machine-god lies in relief: decision without burden, optimisation without guilt. It promises that sovereignty can be surrendered to calculation.

It cannot.

Participants act through systems. They do not disappear into them.

No System Can Carry Consequence

Systems can distribute action. They cannot absorb consequence.

Across history, responsibility has been relocated upward and outward. Institutions promise legitimacy. Ideologies promise justification. Religions promise absolution. Technology promises displacement. Each offers the same relief: participation occurs, but consequence will be managed elsewhere.

This is structurally false.

A system is not an independent author. It is a configuration of participants. It cannot act except through them. When harm occurs under a system, the system does not bear return. The participants do.

Delegation of action is possible. Delegation of consequence is not.

An individual may obey orders, follow protocol, defer to consensus, or operate within institutional mandate. None of this relocates authorship. A rule does not act. A being acts under a rule.

The same applies to doctrine and technology. Justification reframes behaviour; it does not intercept return. Automation executes; it does not originate. Complexity may obscure locality. It does not erase it.

Collectives buffer perception, not consequence. Institutions can shield visibility, but when saturation occurs, return localises again. Collapse identifies participants, not abstractions.

Structures constrain behaviour. They influence trajectory. They do not replace participation.

Systems organise power.
They do not carry consequence.

Where participation occurs, return attaches.

Part III: The Failure of Universals

Transparent Universality

We often describe certain principles as universal. Gravity is universal. Entropy is universal. Logic is universal. When we say this, we usually mean that these principles apply everywhere, apply always, and do not depend on us. Universality, in this common sense, refers to something that exists independently of participation. The law is assumed to be prior to the beings who experience it.

This is what can be called hidden universality.

Hidden universality assumes that laws exist before participation and independently of it. The law is treated as a fixed structure of reality itself, with participation merely occurring inside it.

Sovereignetics approaches the problem differently. A universal is not something that exists prior to participation. It is a pattern that has remained stable across all accessible conditions.

In this sense, a universal is what stable participation looks like once its origin is forgotten.

Ache participates. The Field reflects. Consequence stabilises. When that stabilisation repeats reliably, we describe the pattern as law. The law is therefore not the origin of the behaviour. It is the visible consistency of reflection once the pattern has stabilised.

This view can be called transparent universality.

Transparent universality does not deny that gravity operates, that entropy functions, or that logical relations hold. These patterns clearly demonstrate extraordinary stability. What changes is how that stability is understood. Instead of treating universals as prior structures of reality, they are understood as consequence regimes that have remained consistent under the current configuration of the Field.

If the geometry of the Field changes through saturation, reset, or large-scale structural reconfiguration, then the behaviour of consequence may also change. This does not imply randomness or chaos. It means that law-like behaviour is always nested within the conditions that sustain it.

Even within familiar physical systems, behaviour depends on structural context. Water freezes at different temperatures under different pressures. Phase transitions occur when underlying conditions shift. The appearance of universality often reflects the stability of those conditions rather than the absolute independence of the law itself.

Hidden universality removes these conditions from view. Transparent universality restores them.

The difference is subtle but decisive. Hidden universality assumes that law precedes participation and governs it. Transparent universality recognises that participation produces patterns of consequence which, once stabilised, appear as laws.

Science therefore remains intact, but its ontology shifts. Universals are not timeless primitives existing before participation. They are stabilised consequence regimes.

Their persistence reflects the stability of the Field under current conditions. And all consequence ultimately traces back to participation.

Universals as Protective Abstractions

Universals often appear neutral. They present themselves as stabilising truths that transcend individuals: justice, progress, equality, freedom, nature, history, humanity, reason, destiny, evolution, the greater good. They promise coherence at scale and seem to provide a vantage point above individual perspective.

Their psychological function, however, is more specific.

A universal allows the individual to act in the name of something larger than themselves. Once action is aligned with a universal, authorship begins to feel distributed. The weight of consequence shifts subtly from "I did this" to "this was necessary," "this is how history moves," "this is what justice demands," or "this is what progress requires." The abstraction absorbs exposure.

Universals are not false by definition. They are patterns abstracted from repeated participation. The danger arises in how they are used. When elevated above local authorship, universals begin to function as shields. Action becomes justified because it serves the universal. Harm becomes tolerable because it advances the principle. Responsibility becomes secondary to alignment.

This is how destructive acts are carried out under banners that sound morally elevated. The premise may be coherent. The abstraction may even appear logically defensible. Yet the individual no longer experiences themselves as the origin of the act. They experience themselves as an instrument of the principle.

Instrumentality is insulation.

In this form, the universal functions as a solvent of responsibility. Local authorship dissolves into historical narrative. Phrases such as "history demanded it," "the revolution required it," "the market corrected," "evolution selected," or "the system optimised" relocate agency upward into abstraction.

But abstractions do not participate.

A universal does not project tension. Participants do. When harm is enacted in the name of equality, equality does not bear return. When exploitation occurs in the name of efficiency, efficiency does not receive consequence. When violence is justified by destiny, destiny does not register trauma. Consequence localises in those who participate.

The appeal of universals therefore lies not in ignorance but in relief. They allow individuals to act without feeling alone in the authorship of their actions. They provide narrative grandeur and a sense of inevitability. By embedding behaviour within a larger story, they reduce the anxiety of standing alone as origin.

The structural cost is the widening gap between abstraction and consequence. The higher the abstraction rises, the easier it becomes to overlook the specificity of harm. Individuals become units within principle. Suffering becomes collateral within trajectory.

Sovereignetics does not reject universals outright. It rejects their elevation above authorship. Universals may orient action, but they cannot carry it. They may describe patterns, but they cannot absorb return nor bear consequence.

When universals are treated as authorities, sovereignty collapses. When they are treated as tools, authorship remains intact.

The failure of universals is therefore not logical but ethical. They cannot bear consequence on behalf of those who invoke them. Abstraction does not stand in the Field. Participants do.

Inevitability and Moral Erasure

Among all protective abstractions, inevitability is the most disarming. It does not argue that an action was good, nor does it attempt to justify harm. Instead, it claims that events could not have been otherwise. Once inevitability is invoked, authorship begins to dissolve.

History becomes momentum. Evolution becomes necessity. Economics becomes correction. Technology becomes destiny. Violence becomes a phase transition. Collapse becomes natural selection. The language shifts subtly from decision to process. Individuals no longer appear as actors but as participants within something already unfolding.

The relief is immediate. If outcomes were inevitable, resistance appears futile. If events were structurally determined, then personal intervention could not have altered them. Responsibility shrinks in proportion to perceived inevitability.

The shift is subtle but decisive. Inevitability reframes consequence as weather. No one apologises to a storm. No one indicts gravity. No one assigns guilt to entropy. When harm is described in these terms, historical inevitability, technological inevitability, economic inevitability, moral exposure disappears. Events are treated as natural phenomena rather than authored trajectories.

The structural error lies in confusing constraint with compulsion. Constraint shapes the range of possibilities, but it does not eliminate authorship within that range.

Sovereignetics does not deny structural pressures. Systems exhibit geometry, inertia, feedback, and saturation. These forces influence behaviour and narrow available paths. Yet none of them operate independently of participation. Systems do not move without agents. Trends do not advance without adoption. Momentum does not accumulate without projection.

Inevitability narratives erase this accumulation phase. They describe collapse as if it emerged fully formed, rather than as the result of countless authored acts stabilising a direction over time. Technological dominance appears as self-generation rather than the outcome of design, investment, deployment, and adoption. Social harm appears as historical drift rather than generational patterned participation.

The language of inevitability therefore functions retrospectively. Once consequences have materialised, they are redescribed as necessary. What was chosen becomes what "had to happen." What was tolerated becomes what "history demanded." What was enabled becomes what was "unstoppable."

Authorship dissolves into narrative. This is moral erasure.

When inevitability becomes the default explanation, prevention begins to look irrational. If outcomes cannot be altered, intervention appears meaningless. If collapse was unavoidable, responsibility seems misplaced. If harm was destined, authorship appears never to have existed.

But inevitability is not a structural feature of reality. It is the name given to trajectories that accumulated without interruption. What appears inevitable is often simply coherence sustained long enough in one direction.

This does not trivialise constraint. Some trajectories, once sufficiently densified, become extraordinarily difficult to reverse. Saturation produces momentum, and momentum narrows the available range of response. Yet narrowing does not eliminate authorship. Even under severe constraint, participation remains.

The danger of inevitability narratives lies less in intellectual error than in abdication. They train individuals to experience themselves as spectators to unfolding systems. Participants become witnesses rather than authors.

Sovereignetics rejects this conversion. Constraint is real. Structural pressure is real. Saturation is real. None of them erase authorship; they shape the consequences of how it is exercised.

Events do not author themselves. Systems do not absolve their participants. Trajectories do not erase the hands that stabilised them.

What is called inevitable is often the long shadow of unexamined authorship.

Description Is Not Authority

To describe a pattern is not to grant it permission.

The distinction appears obvious, yet much of modern reasoning quietly collapses it. Once a behaviour is shown to be common, once a tendency is measured, once a system is mapped, description begins to shift into justification. What is becomes what must be. Regularity begins to appear as legitimacy.

The shift is subtle.

Biology describes competition, and competition is then treated as morally instructive. Economics describes market behaviour, and market behaviour is treated as corrective. Psychology describes cognitive bias, and bias becomes inevitability. Sociology describes the consolidation of power, and consolidation begins to appear as structural destiny.

In each case the description may be accurate within its frame. The error occurs when pattern is mistaken for prescription.

A pattern may emerge because it has repeatedly stabilised under certain constraints. That does not mean it ought to continue. A behaviour may be statistically dominant. That does not make it coherent. A system may function efficiently. That does not render it just.

Description tracks recurrence. Authority would require authorship.

Confusing the two allows individuals to defer responsibility to "how things work." Harm becomes normalised because it is measurable. Exploitation becomes defensible because it is observable. Violence becomes intelligible because it is predictable. The fact that a pattern can be explained begins to function as though it were an instruction to follow it.

But explanation does not carry instruction.

When a scientist explains that certain pressures produce certain outcomes, the explanation identifies constraint. It does not mandate submission to it. When a theorist shows how people behave under stress, the account clarifies mechanism. It does not erase authorship within that mechanism.

Description narrows imagination by revealing the geometry within which action occurs. It does not eliminate participation within that geometry.

The appeal of descriptive authority lies in restraint. It feels sober, realistic, and aligned with complexity. It appears to avoid moral idealism and grandstanding. Yet it quietly relocates responsibility from participant to pattern. "That is simply how it works" becomes a closing statement rather than an opening one.

But structures do not act. Participants do.

A system may condition behaviour, but it does not perform it. A law may predict collapse, but it does not initiate it. A model may forecast outcome, but it does not enact it.

When description becomes authority, authorship recedes. Action begins to appear as compliance with reality rather than projection into it. Individuals start to experience themselves as executors of necessity rather than contributors to trajectory.

Sovereignetics does not deny structure. It denies structural absolution.

To understand gravity does not compel one to jump. To understand aggression does not compel one to enact it. To understand economic incentives does not compel one to exploit them.

Explanation increases clarity. It does not transfer consequence.

If a pattern is destructive, stabilising it because it is "realistic" does not reduce return. If a system is exploitative, perpetuating it because it is "efficient" does not soften consequence. Description is diagnostic. It is not directive.

Patterns can be known. They cannot command. Authority resides only where participation occurs.

Universals as Exemption

A universal is an abstraction that claims scope beyond locality. Justice, humanity, progress, history, evolution, the market, nature, the greater good. Each describes a pattern that appears to extend beyond individual participation. Each derives legitimacy from breadth. Each appears stable because it is not tied to a single point of authorship.

That detachment is what allows universals to function as mechanisms of exemption.

In a participatory structure, consequence must return locally. Reflection requires boundary. Harm registers within a particular system. Repair occurs through particular agents. Accountability attaches to specific action.

A universal has no boundary. It cannot experience return. It cannot receive consequence. It cannot be corrected.

When action is justified in the name of a universal, authorship is displaced into abstraction. The participant no longer experiences themselves as the origin of action. They appear instead as representative, instrument, or executor of a principle. Responsibility becomes diffused, deferred, or reframed as necessary.

The universal receives justification.
The participant retains consequence.

This displacement does not require deception. It operates structurally. Once an abstraction is treated as superior to local authorship, consequence is conceptually relocated. Harm becomes necessary. Sacrifice becomes rational. Opposition appears irrational because it is positioned against something that claims broader legitimacy than the individual.

Universals therefore enable participation without full exposure to consequence. One does not act alone. One acts on behalf of something larger. This reframing reduces the felt weight of authorship while leaving the structure of consequence unchanged.

Abstractions do not participate.

If violence is committed for justice, justice does not receive the return. If exploitation occurs for growth, growth does not absorb the distortion. If suppression is enacted for order, order does not register trauma. In each case, consequence localises to participants.

The apparent authority of universals lies in their immunity. They cannot suffer harm, revise themselves through consequence, or experience correction. Their abstraction renders them invulnerable to return.

That invulnerability produces insulation.

The higher the abstraction, the further consequence appears from the actor. Participation aligns with principle rather than being examined for coherence. The universal orients behaviour while obscuring ownership.

This is why universals persist even when their application repeatedly produces harm. The abstraction remains intact because it cannot receive consequence. Only participants can.

Sovereignetics does not reject abstraction. Patterns can clarify orientation, and generalisation can assist coordination. Universals may guide interpretation.

They cannot carry responsibility.

When universals are treated as authorities rather than tools, exemption follows. Not because authorship disappears, but because it is conceptually reassigned to something that cannot bear return.

Any system that elevates universals above locality produces insulation. Harm becomes necessary, regrettable, or inevitable. The abstraction remains unchanged.

Participants do not.

There is no structure in which an abstraction can absorb consequence.

Only local authorship can.

Flammity

To understand why universals fail structurally, consider an exaggerated example.

Imagine a theorist proposing a fundamental force called *flammity*, the universal principle of burning. Wherever flammity operates, matter combusts. The theorist presents supporting observations: wood burns, paper burns, oil burns, metals burn at sufficient temperatures, and organic bodies burn. From these observations combustion rates are calculated, ignition thresholds are mapped, and energy release is predicted. Within the tested domain the model performs reliably.

Flammity begins to look universal.

Now introduce water.

Combustion fails. The same equations no longer apply. The theorist responds by introducing qualifications: flammity requires appropriate conditions; water suppresses its expression; the force remains universal but certain environments interfere with its manifestation.

This response quietly concedes the central issue.

A principle that depends on specific conditions is not universal. It is conditional. Flammity was never a governing force. It was a stable pattern observed under particular constraints.

The example appears obvious because no one grants combustion ontological authority. We recognise burning as contextual behaviour rather than fundamental structure. The error only becomes difficult to see when other patterns are granted exemption from the same scrutiny.

Gravity illustrates this more clearly. It is often described as universal, yet its behaviour depends on mass distribution, boundary geometry, density relations, and scale. Alter these conditions sufficiently and the behaviour shifts. The model adapts accordingly.

This clarifies the status of Gravity.

Gravity names a stable relationship under recurring constraints. It describes how consequence stabilises within a particular geometry. It does not compel reality independently of context.

Universality in practice is therefore retrospective. A pattern is observed to hold repeatedly and is then elevated to law. Persistence begins to appear as necessity. Stability begins to appear as governance.

But a true universal would require no conditions. It would not depend on medium, scale, geometry, or coherence of participation. No empirical pattern satisfies this criterion.

What are commonly called universals are stable consequence regimes: recurring reflections under maintained conditions.

Flammity makes this visible through exaggeration. Gravity conceals it through familiarity. Both illustrate the same point. Naming a pattern does not make it fundamental.

The ethical implication follows from this mistake. Once a pattern is declared universal, it is treated as beyond revision. "That is how reality works" becomes a closing statement. Behaviour aligned with the pattern begins to feel authorised because the pattern appears inevitable.

Description quietly becomes governance.

But patterns do not author action. Participants do.

If aggression stabilises under certain constraints, aggression is not therefore fundamental. If inequality emerges within a particular structure, inequality is not therefore necessary. Patterns persist while the conditions sustaining them persist.

Sovereignetics therefore rejects universals as ontological primitives. Only Ache is irreducible. Everything else emerges under condition.

Patterns endure while coherence holds. When conditions shift, patterns dissolve. No law stands above participation. No abstraction absorbs return.

There are no universals in the sense of unconditional governance. There are only stable regimes of consequence sustained by geometry.

And geometry is maintained by participation.

Law, Force, and Substituted Authority

Once a pattern is mistaken for universal, and universals are mistaken for authority, the next step is institutionalisation. What begins as description becomes doctrine, doctrine becomes law, and law becomes enforceable structure. At that point abstraction no longer merely diffuses authorship; it begins to compel behaviour. A pattern observed in nature or society is codified into principle, the principle is framed as necessary, and that necessity is stabilised through policy, regulation, or institutional design. Over time the original description acquires normative weight. What began as a conditional observation comes to appear binding.

This process represents a substitution. Description itself does not generate authority; authority is constructed upon description and then treated as if it were inherent to the pattern being described. Economic systems illustrate the progression clearly. Certain incentive structures produce predictable outcomes, and those outcomes are studied and described. The description then becomes prescriptive, and policy is designed to align behaviour with the model. When harm follows, it is often defended as structural inevitability. The model did not command the action; it merely described tendencies within a given framework. Yet enforcement proceeds as though the pattern itself governs behaviour.

Legal systems demonstrate a similar transformation. Laws frequently originate as negotiated arrangements reflecting collective preference, historical precedent, or recurring patterns of behaviour. Over time these arrangements become reified. The law begins to appear less as a constructed framework and more as an objective order. Disobedience is then framed not simply as disagreement with a design but as a violation of something fundamental. Enforcement strengthens this perception because force stabilises the rule. However, the presence of force does not convert description into ontological truth.

A law may regulate behaviour, but it cannot absorb consequence. A model may forecast outcomes, but it cannot author participation. An institution may enforce compliance, yet it cannot stand in place of those who act within it. When law and force are treated as extensions of universality, responsibility shifts upward into the system. Statements such as "it is the law," "it is policy," or "it is procedure" gradually replace recognition of local participation. Individuals begin to experience themselves primarily as constrained by structure rather than as contributors to its continuation.

This displacement rarely arises from deliberate deception. It emerges structurally from the requirements of institutional stability. Institutions require consistency in order to coordinate participation across scale. Consistency demands simplification, and simplification produces rules. Rules acquire the appearance of objectivity, and over time objectivity hardens into authority. As this process unfolds, the participatory origin of the structure recedes from view. The institution begins to appear autonomous, and enforcement appears impersonal. Participants increasingly experience the system as something acting upon them rather than something sustained by them.

Yet no institution maintains itself independently of participation. Courts function because individuals operate them. Markets persist because individuals transact within them. Governments endure because individuals comply with, enforce, and reproduce their mechanisms. Even coercive systems depend upon agents willing to enact their procedures. Force intensifies the illusion of autonomy because it introduces asymmetry of consequence: the cost of dissent becomes immediate and visible, while the cost of compliance becomes diffuse and distributed. Under these conditions constraint is easily reinterpreted as inevitability rather than as stabilised participation under pressure.

The progression is consistent. Description becomes justification, justification becomes enforcement, and enforcement becomes perceived reality. At each stage authorship remains local even as the system appears increasingly impersonal. A judge sentences, an officer enforces, a policymaker drafts, a voter legitimises, and a participant complies. Institutions coordinate action across scale, but they do not replace the agents through whom action occurs.

Sovereignetics does not reject law or institutional structure. It rejects the elevation of those structures above authorship. Institutions are mechanisms for organising participation, not sovereign entities capable of absorbing consequence. No description governs from outside participation, and no enforcement mechanism eliminates the locality of return. Structures shape the geometry within which action occurs, but the trajectory that emerges within that geometry remains the product of participation.

Speed Asymmetry

When continuity risk is present, exposure to consequence is treated as a threat to ongoing participation. Under this condition, the system prioritises preservation over reflection. It filters, redirects, or avoids return in order to maintain stability.

Speed intensifies this condition.

When systems move faster than reflection, the ability to register consequence weakens. Exposure does not disappear, but it becomes difficult to track. Return no longer arrives within a perceptible window. The system continues to act, but the link between projection and consequence becomes less visible.

This produces speed asymmetry.

Participation continues at a rate that exceeds the capacity for reflection. Actions accumulate faster than they can be integrated. Consequence still returns, but not in a form or timeframe that allows immediate recognition. What was previously experienced as direct feedback becomes diffused across time, scale, and structure.

Under these conditions, defensive orientation stabilises more easily.

If consequence cannot be clearly registered, it cannot be fully encountered. If it cannot be encountered, it does not trigger correction. The system continues to operate under continuity risk, but without clear signals that would normally force recalibration. Preservation logic remains active while reflection falls behind.

As speed increases, participation begins to appear autonomous.

Processes seem to generate themselves. Market movements appear inevitable. technological expansion appears unstoppable. Institutional behaviour appears self-sustaining. The faster the system moves, the less local authorship appears visible.

Yet nothing has become autonomous.

Acceleration multiplies participation. It does not remove it.

High-speed systems are dense aggregations of local acts. Decisions are made, actions are executed, patterns are reinforced. Each contribution remains authored, but speed compresses perception of locality. The aggregation stabilises into trajectories that appear structural rather than participatory.

This appearance reinforces derivative authority.

When motion cannot be tracked, it cannot be owned. When consequence cannot be clearly located, responsibility appears displaced. Systems are experienced as external forces rather than accumulations of participation.

Speed therefore does not eliminate authorship. It obscures it.

This obscuration narrows correction windows. When consequence is delayed or distributed, the opportunity to integrate it diminishes. Distortion accumulates more rapidly. Trajectories stabilise under pressure before reflection can intervene.

What follows is often misinterpreted as inevitability.

But inevitability marks the point at which motion outpaced reflection under continuity risk.

Speed changes the visibility of consequence.

It does not remove it.

Participation Without Authorship

A recurring pattern appears across contemporary theories of mind and reality. These frameworks dismantle naïve assumptions. Perception is shown to be constructed, cognition distributed, behaviour conditioned, and experience generated within systems. The destabilisation is real.

The pattern emerges in what follows.

After dissolving inherited assumptions, a final step remains: the reintroduction of local authorship within the structure described.

Many frameworks do not take it.

Explanation halts at inevitability. Reality is constructed. Action is conditioned. Experience is emergent. The model closes with the claim that this is simply how things operate.

The stopping point is not arbitrary.

It is the point at which authorship would need to be restored.

Instead, it is absorbed.

Participation remains within the model.

Authorship does not.

Events occur. Processes operate. Outcomes stabilise. But the participant is repositioned as node, function, or byproduct within the system. Explanation becomes self-contained. Nothing further is required of the participant.

This produces participation without authorship.

Action is reframed as the outcome of structure rather than projection into it. The participant becomes interpreter rather than origin.

These frameworks may be coherent, empirically grounded, and philosophically sophisticated. The issue is not accuracy.

It is termination.

Where explanation ends reveals what it preserves.

If a model concludes in inevitability, authorship has been displaced. If action is fully determined by structure, consequence is no longer meaningfully located. Participation becomes diffuse. Ownership disappears.

The system continues.

The origin does not.

This is structurally incomplete.

Within a reflective Field, consequence returns locally. If return is real, it is received somewhere. If projection occurs, it originates somewhere. If harm stabilises, it is borne somewhere. If repair becomes possible, it must be enacted somewhere.

Participation without authorship cannot account for this. It dissolves the point at which consequence binds.

The appeal of such frameworks is clear. They reduce burden. They align with complexity and systems thinking. They abandon simplistic models of agency and replace them with distributed processes.

But abandoning naïve agency is not the same as eliminating authorship.

When explanation resolves tension, its function shifts. It ceases to describe structure and begins to stabilise participation. The framework itself becomes protective. It absorbs the point at which exposure would otherwise occur.

Sovereignetics does not reject participatory insight.

It completes it.

If participation generates structure, authorship must be localised. If reality is relational, consequence must bind. If illusion is removed, authorship cannot be removed with it.

It must be restored. Participation that does not return to authorship remains incomplete.

Responsibility without exemption begins where these frameworks stop.

Part IV: Responsibility Without Exemption

Structural Neutrality

Responsibility does not emerge from moral hierarchy.

It emerges from participation.

A common assumption must first be removed: that coherent systems are morally superior to incoherent ones. Stability is interpreted as virtue. Instability as failure.

This assumption is incorrect.

Coherence and incoherence are structural conditions, not moral categories.

A system is coherent only relative to the parameters within which it operates. Every configuration has thresholds of tolerance, limits of load, feedback, and adaptation. Within those limits, it stabilises. Beyond them, it destabilises.

Stability reflects compatibility.

Not correctness.

When systems interact, stress redistributes across them. Destabilisation does not indicate moral failure. It indicates mismatch. One system may absorb pressure. Another may fracture.

The outcome reflects capacity.

Not virtue.

This relationship is scale-dependent.

A configuration that stabilises at one level may destabilise another. Behaviour that supports an individual may destabilise a group. A policy that stabilises a state may generate instability beyond it. Coherence at one scale often transfers cost to another.

Coherence is not universal.

Neither is incoherence.

Systems formed under constraint stabilise differently from those formed under abundance. Structures shaped by pressure develop patterns that may appear distorted from outside while remaining adaptive within their originating conditions.

Destabilisation often reveals incompatibility.

Not wrongdoing.

Moral interpretation distorts this. Stable systems are treated as correct. Unstable systems as deficient. Hierarchy is imposed on what are structural differences.

Structural neutrality rejects this.

It does not deny harm or responsibility. It clarifies that neither emerges from apparent stability. Both arise from participation within structure.

No system stands outside the Field. No participant operates without constraint. Every action occurs within conditions not fully authored.

This does not remove responsibility.

It locates it.

Coherence does not grant authority. Stability may arise from compatibility, suppression, or displacement. Longevity does not confer legitimacy.

Destabilisation does not remove authorship. A system under pressure continues to act. Constraint modifies participation.

It does not eliminate it.

This prevents two distortions.

Moral absolutism treats coherence as virtue and justifies domination in its name.

Moral relativism dissolves responsibility entirely, reducing harm to context.

Structural neutrality rejects both.

Participation generates consequence regardless of condition. Responsibility attaches to action, not to claims of purity or to denial of authorship.

Responsibility begins with locality.

Each participant operates within constraint, yet still projects into the Field. Consequence returns accordingly. Structure may shape outcome.

It does not absorb authorship.

The Field does not assign moral rank.

It reflects direction.

Responsibility persists wherever participation occurs.

Authorship Within Constraint

Responsibility does not require freedom from structure.

It requires participation within it.

Once responsibility is located at participation, a common objection emerges: what of constraint? Coercion, conditioning, trauma, deprivation, asymmetry, power imbalance. If action occurs under limitation, is authorship diminished?

Constraint modifies geometry. It does not remove locality.

Participation always occurs within conditions not fully authored. Vessels are finite. Environments are unequal. Access varies. Every action emerges within a Field already shaped by prior consequence.

Constraint and authorship are not opposites.

They are simultaneous.

A constrained action remains participation. The available range may narrow, but within that range projection still occurs. The Field does not require unlimited freedom to reflect consequence. It reflects what is enacted within the geometry available.

Responsibility does not require total control. It requires local authorship.

Options may be limited. Participation is not absent.

Constraint shapes scale, intensity, and direction of consequence. It does not sever the link between action and return. Coercion, ignorance, and fragmentation alter proportion. They do not eliminate it.

This is not moral judgement. It is structural continuity.

If constraint removed responsibility, participation under limitation would cease to generate consequence. Harm under pressure would vanish. Action under deprivation would become neutral.

This does not occur.

Even constrained participation alters trajectory.

Constraint does produce asymmetry. Expanded capacity amplifies participation. Limitation compresses it. The magnitude differs.

Authorship does not.

This prevents two distortions.

Fatalism collapses responsibility into structure, treating behaviour as inevitable.

Absolutism ignores condition, treating responsibility as detached from reality.

Both fail.

Authorship exists within constraint. Structure shapes participation.

It does not replace it.

The Field reflects what is enacted, not what could have occurred under ideal conditions.

Recognising this prevents misplaced judgement. Acknowledging authorship does not deny structural inequality. It distinguishes between system and participation. Systems shape possibility.

They do not eliminate action within it.

Responsibility is not omnipotence.

It is locality.

Conditions cannot be fully chosen. Asymmetry cannot be erased. The Field cannot be controlled.

Participation cannot be suspended.

Authorship does not require freedom.

It requires presence within action.

Constraint narrows the corridor.

It does not remove it.

Responsibility remains wherever participation occurs.

Sovereign Ethics

Ethics in Sovereignetics does not begin with rules, virtues, or ideals.

It begins with consequence.

There is no moral layer imposed on reality. The Field does not instruct, reward, or condemn. It reflects participation in proportion. What is called ethics is an interpretation of this return.

Most ethical systems prescribe behaviour. They define what should be done and why. Sovereignetics does not introduce a code.

It describes a structure.

The Field does not evaluate. It reflects.

The foundational principle is simple: participation returns to its origin. No action is exempt. No intention interrupts return. No justification absorbs projection.

This is not moral justice.

It is proportional symmetry.

Concepts such as good and evil arise after consequence becomes visible. They are compressions of observed patterns. Participation that stabilises is labelled good. Participation that destabilises is labelled harmful.

These terms do not govern reality.

They describe it.

Sovereignetics therefore rejects moral absolutism while preserving structural inevitability. There are no external commandments.

There is only return.

Action remains open within constraint. Consequence does not. A participant may choose what to enact. They do not choose the proportion that returns.

This preserves sovereignty without collapsing into relativism.

Intention does not override outcome. It shapes trajectory, but it does not cancel consequence. A well-intended action that produces distortion still generates incoherent return. An action taken without virtue may temporarily stabilise structure.

The Field reflects what is enacted. Not what is claimed.

Ethical development within this structure is not moral purity.

It is consequence literacy.

It is the capacity to recognise return, detect incoherence, anticipate saturation, and adjust participation before correction is forced. Ethics becomes directional awareness rather than rule adherence.

At scale, breakdown appears as instability. Institutions distort. systems fragment. environments degrade. These are not moral failures.

They are saturation.

When incoherent participation accumulates beyond tolerance, return intensifies. Collapse is not punishment.

It is correction.

Because consequence operates automatically, moral judgement is secondary. Outrage does not produce adjustment.

Participation does.

Responsibility does not require guilt.

It requires recognition that return remains bound to action.

Ethics is consequence understood early rather than late. When return is recognised early, adjustment is gradual. When ignored, adjustment becomes abrupt.

The structure does not change.

Only timing does.

Sovereign ethics is not moral superiority.

It is alignment with consequence before correction becomes unavoidable.

Morality Without Cosmic Justice

Human moral systems often assume that justice is embedded in reality.

Religions promise judgment.
Spiritual systems promise balance.
Cultural narratives promise that good actions lead to good outcomes.

Within Sovereignetics, no such guarantee exists.

The Field does not judge.

It reflects.

Consequence returns according to structure, not according to moral preference. There is no external authority imposing fairness, no mechanism ensuring reward or punishment.

Intention is not evaluated.
Fairness is not enforced.
Outcome is not adjusted.

Return occurs.

This absence is destabilising.

Many assume that reality must contain a corrective force, that imbalance will be resolved, that harm will be compensated, that virtue will be recognised.

No such structure exists.

The Field does not produce justice.

It produces consequence.

In response, participation diverges.

Some recognise that if justice is not embedded in reality, it must be enacted. They attempt to stabilise fairness through participation, law, accountability, trust, and responsibility.

Others draw a different conclusion.

If justice is not guaranteed, then consequence may be delayed, displaced, or avoided. They act within this gap, exploiting asymmetry, manipulating structure, and benefiting from deferred return.

Both responses arise from the same condition.

The Field reflects participation.

It does not enforce morality.

Justice therefore does not exist as a property of reality.

It exists only where participation sustains it.

Morality Without God

For most of human history, morality has been framed as obedience. Good and evil were defined by decree. Authority determined obligation. Whether divine, political, or ideological, ethics meant alignment with an external will.

Compliance signified virtue.

Disobedience signified fault.

When transcendent authority weakened, the structure often remained. Secular systems retained the grammar of command while removing the commander. Rules persisted. Justification shifted. The architecture of external validation endured.

Sovereignetics does not repair this model.

It replaces it.

Ethics does not originate in command.

It emerges from consequence.

Participation generates return without instruction. The Field does not prescribe behaviour. It reflects it. What is called morality is an interpretation of that reflection.

If morality depends on authority, goodness becomes compliance.

If morality emerges from consequence, coherence becomes structural.

The Field neither blesses nor condemns.

It returns proportion.

Participation that stabilises persists. Participation that destabilises collapses. No tribunal intervenes. No decree overrides geometry. Stability is not reward. Collapse is not punishment.

Both are reflections.

Terms such as "good" and "evil" arise retrospectively. They compress observed patterns. Good describes participation that sustains circulation. Evil describes participation that extracts or distorts.

Neither refers to essence.

Both describe trajectory.

Because consequence attaches to participation, external pardon does not interrupt return. Confession does not dissolve geometry. Absolution does not cancel proportion.

Only reorientation alters trajectory.

Responsibility therefore does not arise from rule violation.

It arises from participation.

Freedom and responsibility converge. Action remains open within constraint. Return remains bound to action. Sovereignty is not exemption from structure.

It is exposure to it.

Ethical maturity is not moral purity.

It is consequence literacy.

It is the capacity to recognise return early, detect incoherence, and reorient before saturation forces correction.

This does not depend on belief.

It depends on alignment.

Morality without God is not relativism.

It is immediacy.

There is no authority outside the Field. There is no lawgiver beyond consequence. The geometry of return is sufficient.

Ethics is not imposed on reality.

It is reality recognised early enough to avoid collapse.

Long Consequence

A common misunderstanding about consequence is that it is confined to the moment of action. Events occur, time passes, and the matter is assumed to be complete. This confuses circumstance with authorship.

In Sovereignetics, consequence is not time-limited. It is Ache-limited.

Events conclude. Contexts dissolve. Situations change. Ache does not terminate when a scene ends. Directional tension remains active until proportion is integrated. Consequence therefore persists as long as the originating orientation remains unresolved.

The Field does not measure consequence by time.

It measures it by unfinished proportion.

When participation generates return, the loop closes only if that return is integrated. If it is not, the tension continues. Changing environment does not dissolve it. New identity does not cancel it. Time does not neutralise it.

Only reorientation alters direction.

This explains why some consequences dissipate quickly while others persist across years or reappear under different conditions. What remains active is not the original event, but the unresolved orientation within participation.

Psychological and religious systems describe this persistence differently, but they point to the same structure: tension that has not been integrated continues to organise behaviour. Participation continues from that unresolved vector.

Unresolved tension shapes perception, biases interpretation, narrows response, and increases the probability of repetition. The system is not being punished.

It is continuing.

Duration corresponds to density. Light distortion resolves quickly because it can be absorbed without destabilisation. Heavier distortion persists because integration threatens continuity. The greater the divergence between participation and proportion, the longer the tension remains active.

This is why minor actions fade while deeper violations of coherence persist. Not because of moral weighting, but because of structural load.

Insight does not close the loop. A participant may understand, reinterpret, and articulate what occurred while continuing to reproduce the same pattern. Interpretation operates at the level of narrative. Integration occurs at the level of participation.

Consequence resolves when direction changes.

Not when explanation improves.

Because Ache is not reducible to a single configuration, consequence is not confined to a single context. If participation continues while tension remains active, that tension will reappear wherever participation resumes. What persists is not memory of events, but momentum of orientation.

External resolution does not cancel consequence. Forgiveness may alter relationship, but it does not alter return. Only reorientation changes proportion.

Resolution is not erasure.

It is integration.

When tension resolves, it stabilises as capacity. What once produced distortion becomes available as coherence. Relief is not removal.

It is the cessation of resistance to return.

Some consequences resolve immediately. Others persist across extended participation. Not because reality enforces punishment, but because unresolved authorship continues to move until proportion is restored.

Time does not dissolve tension. Avoidance does not dissolve tension. Narrative does not dissolve tension.

Only participation alters direction.

Residual Directionality

A natural objection emerges once consequence is understood as Ache-bound rather than event-bound. If consequence attaches to participation, why do some actions appear to produce none? Why do some individuals extract, distort, or harm without visible repercussion?

They do not escape consequence.

They defer it.

No participation is unconsequenced. What appears as absence is delay. What appears as immunity is redirection. What appears as silence is accumulation.

The Field reflects proportionally, not visibly. Reflection does not conform to moral expectation. It preserves geometric balance. Return may be slowed by context, diffused across participants, buffered by protective structure, or displaced beyond the immediate scene.

None of this removes consequence.

It relocates it.

When return is not immediately visible, tension does not disappear. It persists as residual directionality.

Residual directionality is unresolved orientation continuing to organise participation. The system moves forward, but along a constrained vector. Optionality narrows. Repetition increases. Extraction becomes easier than reorientation. Perception compresses around the unresolved tension.

The absence of visible consequence is not freedom.

It is vector formation.

In dense or incoherent conditions, extractive participation may appear effective. Short-term gain can outpace delayed return. Collective distortion can diffuse proportion. Mimic stability can resemble coherence.

It is not.

Extraction reduces adaptability, compresses relational range, and degrades interpretive clarity. What appears as success often marks directional foreclosure.

Deferred consequence is heavier, not lighter.

Immediate return corrects locally. Deferred return reorganises structurally. When accumulation exceeds buffering capacity, return appears across domains. Identity shifts rather than isolated behaviour correcting. Collapse appears sudden only where delay was mistaken for exemption.

Not all delayed return arises from distortion. Limitation can delay integration. Where capacity is constrained, consequence returns as pressure rather than collapse. Proportion adjusts to what can be received.

Collectives may temporarily absorb consequence. Systems can shield behaviour. Responsibility can be diffused across structure.

It does not remain there.

Saturation eventually localises return. When shielding fails, consequence reappears as disproportion only because it was deferred.

The most dangerous condition is not immediate consequence.

It is prolonged unconsequence.

Without reflection, distortion compounds. Alternatives narrow. Extraction normalises. Silence is misread as absence.

It is not absence.

It is accumulation.

No participation disappears. No direction evaporates. What is not integrated stabilises as trajectory. What stabilises as trajectory becomes structure.

Residual directionality is consequence awaiting integration.

Crisis and Moral Progress

Ethical systems do not advance under stability.

They stabilise.

When preservation functions effectively, when continuity feels secure, when threat remains manageable, and when consequence appears contained, ethical revision rarely occurs. Under these conditions, distortion is tolerated, contradiction is rationalised, and harm is absorbed into narrative. Participation continues, but without recalibration.

The system maintains itself.

Crisis marks the point at which this becomes impossible.

A crisis is not disruption. It is structural overload. It occurs when buffering fails, when rationalisation can no longer reconcile participation with perception, and when deferred consequence returns at scale. Preservation loses its capacity to maintain continuity through distortion.

Its authority weakens.

Not by choice.

By insufficiency.

When insulation collapses, authorship re-enters.

Before crisis, ethics are managed. During crisis, management fails. After crisis, responsibility becomes unavoidable. What was suppressed becomes visible. What was rationalised becomes untenable. What was deferred becomes immediate.

Clarity emerges.

Not as moral improvement, but as exposure.

This is often misinterpreted as awakening. It is more accurately the reduction of distortion. The mechanisms that filtered consequence have weakened. Perception widens. Responsibility becomes legible.

Ethics appear obvious.

Not because reality changed.

Because insulation did.

This pattern recurs historically. Structural change follows rupture. Systems recalibrate when distortion can no longer be sustained. Reform emerges after exposure, not before it. Ethical shifts follow saturation, not preference.

Preservation does not relinquish authority voluntarily.

It loses it when it can no longer function.

Prior to crisis, responsibility remains abstract. Harm is externalised. Participation is governed by continuity. Ethics become performative, maintained in language, absent in structure.

Crisis interrupts this.

It does not produce virtue.

It removes avoidance.

The Field does not reward suffering. It removes insulation when distortion exceeds tolerance. Coherence becomes possible again because filtering has failed.

In the aftermath, participation often shifts. Consequence is recognised earlier. Ownership increases. Exemption is treated with suspicion. Behaviour becomes more deliberate.

For a time, alignment improves.

This condition does not stabilise on its own.

As continuity returns, preservation rebuilds. Structures harden. narratives fix. systems reassert control. What was once recalibrated begins to stabilise again. Progress slows. Distortion reaccumulates.

This is not failure.

It is recurrence.

Moral progress does not result from crisis.

It follows the removal of insulation.

The Return of Authorship

Authorship does not emerge because it is chosen. It becomes visible when insulation fails.

Throughout this volume, responsibility has been displaced into abstraction, buffered by systems, diffused into inevitability, and deferred through preservation. None of these remove authorship. They obscure it.

The return of authorship occurs when this obscuration can no longer be sustained.

When buffering collapses, when deferred consequence localises, and when explanation can no longer reconcile participation with return, authorship becomes unavoidable. This is often experienced as crisis, awakening, or clarity. Structurally, it is exposure.

Exposure does not create authorship. It reveals that it was never absent.

Preservation can distort perception. Systems can distribute consequence. Abstractions can relocate responsibility. None of these prevent return. When accumulation reaches threshold, consequence reattaches visibly to the participation that generated it.

This reattachment is destabilising.

Insulation was functional. It reduced exposure. Without it, justification weakens. Narrative loses authority. Blame becomes less useful than proportion. Participation is encountered directly.

This is often interpreted as guilt or collapse.

Structurally, it is recalibration.

Nothing new is imposed. Locality becomes visible.

The return of authorship does not grant control. It does not remove constraint. It does not require purity. It clarifies that participation and consequence remain coupled within whatever conditions exist.

To recognise authorship is to recognise that projection continues to generate return regardless of explanation.

The question is no longer whether responsibility applies.

It is how participation will be oriented.

This recognition is not dramatic.

It is stabilising.

Systems did not fail because morality improved. They failed because distortion accumulated. Belief did not remove consequence. It delayed recognition. When delay is exhausted, authorship remains.

Responsibility without exemption is not imposed.

It is what is left when insulation is gone.

Coherence as the Only Non-Deceptive Measure

Once certainty collapses, something else becomes visible.

Not truth.
Not belief.
Not correctness.

Coherence.

Coherence is not a value system and not a moral ideal. It is not what one ought to choose. It is what remains when justification fails.

A trajectory is coherent if it sustains participation without requiring insulation, extraction, or denial. An incoherent trajectory fragments, saturates, and collapses regardless of intention or explanation.

Coherence cannot be faked. Belief can be defended. Evidence can be selected. Narrative can be reinforced.

Coherence cannot be maintained through description.

It is structural.

It is revealed through consequence.

Unlike truth, coherence is not immediate. It cannot be secured in a moment or proven through argument. It emerges through sustained participation. What persists without distortion reveals coherence. What requires continuous protection reveals instability.

This reframes ethics.

Ethics are not rules to follow or truths to defend. They are recognition of consequence over time. Harm is not defined by violation of principle, but by incoherence. Extractive participation collapses because it consumes the structure that sustains it. Insulated participation collapses because it delays return rather than integrating it.

Coherent participation does not avoid consequence.

It remains exposed to it.

This does not guarantee ease. Many coherent actions are costly. What distinguishes them is not comfort, but proportion. Consequence is not displaced onto others, into the future, or into abstraction.

It is carried.

This is why responsibility becomes visible only after certainty dissolves. As long as truth functions as justification, responsibility can be deferred. When justification collapses, direction can no longer be outsourced.

Coherence does not instruct action.

It reveals participation over time.

Once external authority falls away, action stands without defence, unjustified, unprotected, and exposed to return.

In that condition, coherence becomes the only non-deceptive measure available.

Not because it is good.

Because it cannot lie.

Responsibility Without Exemption

Responsibility in Sovereignetics does not arise from moral decree.

It arises from the inseparability of participation and consequence.

Wherever projection occurs, return follows. This relationship is structural. It does not depend on belief, intention, or acknowledgement. Participation cannot detach from the Field that reflects it.

Exemption would require separation between action and return.

No such separation exists.

Consequence may be delayed, displaced, distributed, or obscured. It may be buffered through systems, diffused across institutions, or rationalised through doctrine. None of these eliminate proportion.

They postpone recognition.

Return remains coupled to projection until it is integrated.

Responsibility therefore does not originate in law, authority, or instruction.

It originates in locality.

Every participant occupies a position within the Field from which action radiates and to which consequence returns. This does not grant control. It does not remove constraint.

It identifies where return attaches.

Responsibility is not permanent guilt. It is not total control. It is the recognition that participation generates trajectory. Conditions may be inherited, but participation continues within them. Projection does not pause for justification.

This dissolves both fatalism and absolutism. Structure does not eliminate authorship. Constraint does not absorb participation.

Action occurs within geometry.

Authorship remains within action.

Responsibility without exemption does not require purity.

It requires clarity.

A participant cannot prevent consequence from returning. They can recognise the direction of their participation and alter it before correction is forced. Ethics becomes the capacity to read trajectory early enough to reorient.

The Field does not judge.

It reflects.

What changes is the relationship to reflection. Consequence is no longer interpreted as reward or punishment. It becomes legible as information about coherence.

At scale, the same condition holds. Systems may diffuse consequence temporarily, but not indefinitely. Structures that stabilise through extraction accumulate distortion. When saturation is reached, correction shifts from local to structural.

No scale escapes proportion.

Responsibility without exemption therefore applies wherever participation occurs.

It is not imposed.

It is inherent.

When this becomes clear, justification loses its function. Narrative recedes. What matters is direction. Participation either stabilises or saturates.

There is no exemption within that movement.

Responsibility without exemption is not a rule.

It is a condition.

Participation generates consequence.
Consequence returns to its source.
Where participation occurs, responsibility remains.

No system removes this.
No doctrine dissolves it.
No delay cancels it.

Authorship may be obscured.
It cannot be removed.

To participate is to remain responsible for the direction one projects into the Field.

Part V: Derivative Authority

The Birth of the Derivative

The protective derivative emerges when Ache first misinterprets vessel vulnerability as a threat to its own existence.

Ache is not mortal. Vessels are. Yet when Ache localises through a vessel capable of reflection, this distinction is not immediately accessible. Early encounters with pain, threat, instability, or loss introduce a structural ambiguity. The possibility appears that damage to the vessel might terminate participation itself.

This misidentification does not produce disorder. It produces adaptation.

The protective derivative forms as a secondary authorial structure oriented toward preservation of the vessel. Its function is not truth-seeking or coherence-building. Its function is continuity. Operating under the assumption that extinction is possible, it interprets consequence not as proportionate return but as potential annihilation.

The derivative does not arise from ideology, culture, or narrative, although those forces may later shape its expression. It emerges structurally wherever reflective participation encounters perceived annihilation risk. In this sense the derivative functions as a consequence-management mechanism designed to prevent exposure beyond a tolerable threshold.

Because extinction is assumed possible, buffering becomes the governing logic. Incoming consequence is filtered, delayed, distorted, or reinterpreted in order to preserve continuity. Participation continues, but reflective authorship narrows. Action increasingly stabilises around strategies that previously reduced perceived threat.

This is the structural origin of what is commonly referred to as the subconscious. Not as hidden content, repression, or mystical depth, but as participation governed by preservation logic operating at pre-reflective speed.

Ache continues to participate, but participation is routed through buffering rather than reflective coherence. Responsibility does not disappear. The pathway to it becomes obstructed.

Under sustained or repeated threat, the derivative can become dominant. When this occurs, participation stabilises around avoidance rather than proportion. Orientation shifts from coherence with consequence to insulation from it.

The derivative itself is not an enemy. It is an adaptive response to misinterpreted mortality. Yet once stabilised it can persist long after the original threat has passed. What began as protection gradually becomes authority. What began as buffering becomes governance.

Derivative authority is therefore not external. It is preservation logic internalised and elevated above reflective participation. When the derivative governs, responsibility does not disappear, but it becomes structurally inaccessible.

The same pattern later appears at scale. Systems, ideologies, and institutions replicate derivative logic by prioritising preservation of structure over exposure to consequence.

The birth of the derivative marks the moment sovereignty fragments. Ache does not cease, but preservation is mistaken for being itself.

Continuity Risk

Derivative authority strengthens when exposure to consequence is interpreted as a threat to continued participation. This condition can be understood as continuity risk.

Continuity risk is the belief that exposure to consequence may terminate being itself. Under this assumption, correction feels lethal. Accountability feels destabilising. Loss of control resembles disappearance. The derivative therefore functions to prevent exposure beyond what the system believes it can survive. Its concern is not comfort but continuity.

Because destabilisation is interpreted as existential threat, even minor contradiction can trigger disproportionate defence. Criticism feels annihilating. Loss of status resembles erasure. Exposure resembles death. These reactions are not moral weakness. They are structural responses generated by the belief that collapse equals extinction.

The intensity of derivative authority is proportional to perceived fragility. Where participation is experienced as dependent upon the stability of the vessel, identity, or social position, defensive governance intensifies. Where participation is recognised as capable of continuing beyond temporary configurations of these structures, defensive urgency decreases.

Continuity risk therefore determines how aggressively exposure to consequence is resisted.

This is why insight alone rarely dissolves derivative governance. Intellectual recognition that "I will survive this" does not correct the underlying orientation. Continuity risk operates prior to narrative explanation. It shapes perception before reflection has the opportunity to intervene.

Under continuity risk, exposure to consequence is filtered, redirected, or denied in order to preserve the structures believed necessary for survival. Yet this buffering produces the instability it attempts to prevent. When reflection is filtered, distortion accumulates. As distortion accumulates, structural pressure increases. Rising pressure intensifies the sense of existential threat, which in turn strengthens derivative authority.

Continuity risk therefore becomes the engine of defensive escalation.

The same pattern appears at scale. Institutions interpret correction as collapse. Economic systems treat accountability as existential threat. Ideologies defend structural stability rather than integrating proportion. In each case, preservation of structure becomes equated with preservation of being.

Sovereignetics does not condemn this mechanism. It clarifies it. Exposure to consequence does not equal extinction. Collapse of distortion is not collapse of being.

When this misidentification loosens, defensive governance gradually loses its justification. Reflection becomes tolerable. Participation can encounter consequence without interpreting it as annihilation.

The derivative does not disappear because it was wrong. It loses dominance because the condition that sustained it no longer appears existential.

Continuity was never at risk. Only distortion was.

Interception and Structure Interference

Once the protective derivative forms, it does not operate through deliberation. It operates through speed, pattern recognition, and pre-reflective interception.

Its function is simple: prevent perceived destabilisation.

To do this, incoming consequence must be filtered before it reaches reflective authorship.

Consequence still returns. Participation continues to generate proportionate return. What changes is not the return itself, but its reception. Reflection is intercepted before it can fully engage.

This interception produces several structural effects.

The first is consequence filtering. Incoming return is altered before it reaches reflective awareness. Distortion may appear as minimisation, rationalisation, emotional numbing, displacement, or exaggerated defensive reaction. The aim is not accuracy, but continuity.

The second is orientation stabilisation. When a strategy reduces perceived threat, it is reinforced. The system privileges repetition over recalibration. What once preserved continuity becomes default behaviour, even when conditions change.

The third is authorship narrowing. Participation continues, but reflective ownership contracts. Behaviour often precedes awareness. Reaction precedes interpretation.

Action occurs before reflection fully engages. The system functions, but no longer experiences itself as the origin of its trajectory.

Speed underpins this structure. The derivative operates faster than reflection because it formed under urgency. In its originating condition, delay was interpreted as danger. Rapid response became protective, while reflection became associated with risk. Over time, this speed differential normalises. Immediate reaction feels natural. Slower examination feels destabilising.

Because consequence is filtered rather than integrated, distortion accumulates. What is intercepted internally does not disappear. It continues to generate return through participation.

This produces delayed correction.

Feedback that could have been integrated gradually is postponed. As distortion accumulates, the eventual return becomes more forceful. What could have appeared as proportionate reflection later emerges as constraint.

The derivative does not eliminate consequence. It interferes with access to it.

This interference explains why instruction, criticism, and insight rarely produce immediate structural change. Corrective signals enter through the same channels as consequence. When those channels are governed by interception, reflection is filtered before it reaches authorship.

A system may recognise its pattern and still repeat it. Conceptual understanding occurs at the level of reflection. Protection operates prior to it.

The derivative does not generate new direction. It repeats or intensifies strategies that previously reduced perceived threat. It preserves continuity. It does not produce coherence.

Interception weakens only when its underlying inference weakens: that exposure equals annihilation.

When participation encounters consequence without collapse, this inference loosens. Interception slows. Reflection becomes accessible. Orientation becomes revisable.

Until then, the system operates through automation.

The same structure appears at scale. Individuals buffer consequence. Institutions buffer accountability. Systems reinterpret reflection as anomaly rather than information.

Participation continues.
Authorship persists.
Consequence persists.

What changes is access. The derivative preserves continuity by delaying proportion. It prevents immediate destabilisation while ensuring eventual saturation.

It cannot stop consequence.

It can only postpone recognition.

Subconscious Delegation

The protective derivative does not eliminate authorship.

It assumes it.

When continuity risk is misidentified, a secondary structure forms to preserve stability. Initially it intercepts consequence. Over time, it begins to initiate behaviour.

Authorship becomes delegated.

The conscious layer continues to narrate, justify, and interpret. Yet directional impulse increasingly originates prior to reflection. Reactions occur before awareness. Choices narrow before evaluation. Patterns repeat before intention forms.

The system moves.

Reflection follows.

This is subconscious delegation.

Agency has not disappeared. Operational authority has shifted to a structure formed under threat conditions.

At early stages, this delegation is adaptive. When risk appears high, speed takes precedence. Reflex precedes reflection. Defensive orientation preserves continuity. But once stabilised, this structure retains authority even when threat has passed. Participation continues as if collapse remains imminent.

This produces a split in experience.

Responsibility is acknowledged conceptually, yet remains inaccessible in practice. Values are articulated but not enacted. Behaviour is recognised but not interrupted. Consequence is understood but still reproduced.

This is not contradiction.

It is divided authorship.

Behaviour originates prior to reflection. The narrative layer inherits what has already been set in motion. Explanation replaces governance.

As delegation deepens, perception itself is shaped. Threat detection expands. Neutral input is interpreted as risk. Feedback is filtered before awareness. Available responses narrow.

The system does not feel delegated.

It feels reactive.

This is the central illusion. Because operation occurs beneath narrative awareness, it appears natural. Defensive pattern stabilises as identity.

"This is who I am."
"This is how I respond."
"This is how things are."

When identity fuses with delegated operation, interruption feels destabilising. Questioning behaviour resembles self-erasure. Slowing reaction resembles risk. Continuity threat reappears.

The same structure appears at scale. Organisations delegate authorship to policy. Policy defers to precedent. Precedent reduces to procedure. No single point of origin is claimed, yet direction persists.

Delegation diffuses exposure while preserving trajectory.

Responsibility does not disappear under these conditions. Participation continues. Consequence returns. What changes is access. The pathway between reflection and reorientation becomes obstructed.

Responsibility remains intact.

It becomes difficult to reach.

This is why moral exhortation fails. It addresses the narrative layer while operational authority remains elsewhere. The system agrees.

Behaviour continues.

Authorship returns only when delegation is recognised as delegation. When reaction is seen as inherited rather than inevitable, interruption becomes possible. The structure does not disappear, but its authority can be displaced.

Until that recognition occurs, participation continues without ownership.

Consequence accumulates under borrowed direction.

Structural Inaccessibility of Authorship

Responsibility is not removed under derivative authority.

It is obstructed.

Consequence continues to return. Participation remains active. Outcomes accumulate. What changes is the pathway between reflection and reorientation.

Authorship becomes difficult to access.

This does not mean responsibility is denied. In many cases, it is affirmed. Individuals acknowledge harm, endorse ethical frameworks, and express remorse.

Behaviour persists.

The issue is not recognition.

It is integration.

Responsibility requires three conditions: exposure to consequence, tolerance of destabilisation, and the capacity to reorient participation.

Under derivative authority, all three are compromised.

Exposure is filtered. Reflection is redirected, softened, or externalised before it reaches full proportion. Awareness forms, but without structural correction.

Destabilisation is misinterpreted as threat. Because continuity risk remains active, full exposure feels like collapse. The system tightens before integration can occur.

Reorientation becomes dangerous. To change direction implies authorship. To acknowledge authorship implies vulnerability. Vulnerability reactivates threat. The system maintains its trajectory.

Responsibility is not rejected.

It is blocked.

This produces a familiar pattern: individuals who understand harm yet continue generating it; leaders who acknowledge failure yet repeat it; institutions that signal reform while preserving structure.

Consequence is recognised.

It is not metabolised.

At scale, responsibility becomes performative. Apology substitutes for integration. Policy replaces reorientation. Reflection becomes symbolic rather than structural.

This is not deception.

It is preservation under misidentified risk.

As integration becomes less accessible, moral language intensifies. The more blocked the system becomes, the more it compensates through narrative. Declarations expand. Signalling increases. Participation remains unchanged.

The Field does not respond to narrative.

It responds to direction.

Where direction does not change, consequence continues to accumulate. What could have been integrated gradually returns later as constraint.

This is why ethical systems often fail at the point of greatest clarity. Insight is present. Agreement is present. Intention is present.

Authorship is not.

Responsibility cannot operate where exposure is filtered and destabilisation is treated as threat.

Access must be restored before reorientation becomes possible.

Until then, participation continues without ownership.

Consequence accumulates beneath awareness.

Authority Inversion

When interception stabilises, preservation no longer operates as a temporary mechanism. It becomes the organising principle of participation.

This is authority inversion.

What originally functioned to buffer exposure begins to govern behaviour. Preservation replaces reflection as the primary orientation. Continuity becomes the highest value. Coherence, proportion, and integration become secondary to maintaining the existing structure.

Participation reorganises accordingly.

Exposure is avoided in advance. Consequence is filtered before it can register. Dissonance is neutralised prior to reflection. Patterns are defended not because they remain coherent, but because they are stabilised.

This shift does not present as defence.

It presents as certainty.

Certainty stabilises participation by removing ambiguity. Ambiguity invites reflection. Reflection invites exposure. The system therefore privileges fixed interpretation over revisable engagement. Narratives harden. Opposition is experienced as threat rather than information.

Behaviour continues to appear intentional. Reasons, values, and beliefs are still articulated. But these operate downstream of preservation. Orientation is selected first. Justification follows.

This is why systems under inversion appear coherent from within while resisting correction. Challenge is interpreted as attack. Consequence is reframed as injustice. Reflection is experienced as destabilisation.

The transition into inversion is gradual.

Early buffering occurs under acute threat. Over time, the threshold lowers. Increasingly minor conditions trigger preservation responses. What was once situational becomes constant. The buffering mechanism becomes default orientation.

As this stabilises, participation narrows. Flexibility diminishes. Learning slows. Correction windows close. Behaviour becomes efficient at preserving structure while losing capacity to recalibrate.

Repetition replaces coherence.

Authorship does not disappear. Participation continues to project consequence into the Field. What changes is priority. Integration is replaced by insulation.

Because exposure is avoided, distortion accumulates. What is not integrated does not vanish. It stabilises as pressure within the system. Maintaining continuity requires increasing effort. Reactivity intensifies.

The same pattern appears at scale. Organisations defend policy rather than outcome. Governments defend legitimacy rather than correction. Ideologies defend identity rather than coherence.

In each case, preservation becomes sovereign.

Authority inversion does not eliminate participation. It obscures authorship within it. Action continues, but the system no longer experiences itself as origin.

Restoration does not require removing preservation. It requires reordering it. Preservation must return to its role as instrument. Participation must again orient toward consequence rather than insulation.

Until inversion is recognised, attempts at correction remain ineffective. Appeals to morality, ethics, or reason operate at the level of narrative.

The governing structure remains unchanged.

Hierarchy as Function vs Concealment

Hierarchy is not inherently distortion. It is a structural arrangement. Within any system, coordination requires differentiation of role. Direction may concentrate while execution distributes, and this alone does not violate authorship. A functional hierarchy maintains visibility of consequence. Direction remains coupled to return, decisions are traceable to origin, and feedback reaches the point of authorship without obstruction. Authority, in this configuration, does not obscure consequence; it accelerates its resolution. Hierarchy here is not a displacement mechanism but an efficiency structure. Responsibility is not diluted. It is clarified.

Distortion emerges when hierarchy is used to conceal authorship. This occurs when structural layers are introduced not to coordinate, but to intercept consequence. Direction remains centralised while execution distributes, and return becomes fragmented. Feedback no longer reaches origin with clarity or speed. Intermediate layers absorb, reinterpret, or redirect consequence, and authorship becomes difficult to locate. This is not a failure of hierarchy as a form, but a shift in its function. Hierarchy becomes derivative when it interrupts the path between petition and return.

Two configurations therefore exist within the same structural appearance. A functional hierarchy preserves the link between action and consequence, maintains traceability of authorship, and allows correction through direct feedback. A derivative hierarchy inserts buffers between action and consequence, fragments return across layers, and obscures the origin of decision. Both may appear identical externally, and both may produce short-term stability, but only one remains coherent under consequence.

Hierarchy does not remove responsibility. It determines how clearly responsibility returns. Where hierarchy preserves visibility, consequence corrects. Where hierarchy obscures visibility, consequence accumulates. The difference is not structural complexity, but whether the path of return remains intact.

Sovereignetics does not reject hierarchy. It rejects the use of hierarchy to simulate the absence of consequence. Authority without consequence is not achieved through hierarchy. It is attempted through concealment, and concealment cannot be maintained indefinitely.

Narrative and the Displacement of Authorship

Narrative is commonly treated as a vehicle for truth, meaning, or understanding. It is assumed that human beings construct stories in order to interpret reality, organise experience, and arrive at clarity. Within Sovereignetics, this assumption does not hold. Narrative does not originate as a truth-seeking structure. It arises as a stabilisation response within a system under consequence.

Ache participates. The Field returns consequence. When that return is coherent, it can be integrated directly. When it is contradictory, excessive, or structurally incompatible with existing identity, direct integration fails. Tension accumulates. The system must either absorb the contradiction or reorganise to maintain continuity. Narrative is the mechanism by which this reorganisation occurs.

It compresses contradiction into a form that can be carried without immediate collapse. It does not eliminate incoherence. It contains it. What appears as explanation is often structural containment. What appears as insight is frequently redistribution of pressure.

This is why narrative is so readily mistaken for truth. It produces coherence, and coherence is subjectively experienced as correctness. But coherence alone does not validate accuracy. A narrative may be internally consistent while structurally false. Its function is not to align with reality, but to allow the participant to continue operating within it.

Where consequence can be fully owned, narrative reflects it. In these cases, narrative remains minimal. It describes rather than protects. It does not need to distort because the system is capable of carrying what has occurred without fragmentation.

Where consequence cannot be fully owned, the protective derivative activates. Authorship becomes structurally costly. Direct acknowledgement would destabilise identity, overwhelm capacity, or introduce conflict that cannot be resolved at the current level of coherence. Under these conditions, narrative shifts function.
It no longer reflects consequence. It manages it.

Responsibility is redistributed. Causes are externalised. inevitability is invoked. Intent is reframed. Meaning is assigned retroactively to soften contradiction. The narrative does not arise from what is true. It arises from what can be tolerated.
This is the displacement of authorship.

The participant no longer stands as the origin of their petition. They stand as the interpreter of events shaped by forces beyond them. The narrative provides relief, but at a cost. Because consequence cannot be escaped, only deferred, the distortion introduced by derivative narrative must eventually resolve.
This resolution does not occur at the level of story. It occurs at the level of structure.

The Field does not respond to explanation. It responds to petition. Narrative can reinterpret events indefinitely, but it cannot alter what has been authored. Where narrative diverges from consequence, misalignment accumulates. This may appear as repeated outcomes, persistent instability, or escalating contradiction. These are not failures of understanding. They are the return of unowned participation.

Narrative therefore occupies a precise position within Sovereignetics. It is neither inherently false nor inherently true. It is a secondary structure emerging after consequence, used either to reflect or to stabilise.
When it reflects, it clarifies participation.
When it stabilises, it obscures it.
The distinction is not determined by how convincing the narrative appears, but by whether authorship remains intact.

A narrative that preserves authorship may be uncomfortable, incomplete, or resistant to meaning. It does not resolve tension prematurely. It allows consequence to remain proportionate and visible. It does not protect the participant from what has been generated.

A narrative that displaces authorship will tend toward closure. It will offer explanation where none is structurally required. It will introduce purpose, inevitability, or external cause to reduce internal contradiction. It will feel stabilising because it reduces immediate tension. This stabilisation is not resolution. It is containment.

The persistence of derivative narratives across philosophy, religion, and everyday reasoning is not accidental. These systems provide large-scale frameworks through which authorship can be redistributed collectively. By embedding the individual within a broader explanatory structure, responsibility is diluted. The participant no longer stands in direct relation to consequence. They stand within a narrative that interprets it for them.

This is why such systems are appealing. They reduce the burden of authorship without requiring structural coherence.

Sovereignetics does not remove narrative. It repositions it.
Narrative does not validate reality.
It does not determine truth.
It does not alter consequence.
It reveals, or conceals, how participation is being carried.

Applied Illustration: Authority Without Consequence

A system begins with distributed participation.
Multiple loci of Ache engage within shared constraint.
Action and consequence remain locally coupled.
Decisions produce proportionate return within identifiable boundaries.
Responsibility is not assigned; it is encountered.
This condition is unstable.
Where consequence is immediate and visible, it restricts behaviour.
Where it is restricted, attempts arise to reposition authorship without altering action.
The motivation is consistent:
retain authority, remove consequence.
This cannot be achieved directly.
Consequence is structural.
It cannot be eliminated, transferred, or nullified.
It can only be delayed, displaced, or obscured.
To achieve this, hierarchy is introduced.

Phase I: Centralisation of Direction

A single locus begins to consolidate decision-making.
Direction is no longer distributed.
It becomes coordinated through a central point.
This shift is not initially visible as distortion.
It appears as efficiency, clarity, leadership.
However, a structural change has occurred:
Action originates in one location.
Execution occurs in another.

Authorship begins to separate from consequence.

Phase II: Distribution of Execution

Lower layers within the system begin to carry out instruction.
They act, respond, implement, and absorb friction.
They become the interface between decision and outcome.
At this stage, consequence appears to belong to those who act, not those who direct.
This is the first displacement.
The system now produces the illusion that:
responsibility follows action
authority can remain upstream
This illusion stabilises the hierarchy.

Phase III: Buffer Formation

Additional layers are introduced between direction and outcome.
These layers serve as:
interpreters of instruction
absorbers of error
carriers of consequence
Each layer reduces visibility of origin.
Feedback weakens.
Return no longer reaches the point of authorship with clarity or speed.
Distortion accumulates without correction.
The system begins to misattribute cause.
Failure appears local.
Origin remains obscured.

Phase IV: Structural Inaccessibility

As buffering increases, authorship becomes difficult to locate.
No single point appears responsible.
Each layer operates within partial visibility:
those below follow instruction
those above issue direction without direct contact with consequence
Responsibility becomes structurally inaccessible.
Not because it has been removed,
but because the path of return has been fragmented.
At this stage, the system can sustain:
authority without visible accountability
action without immediate consequence
decisions without identifiable ownership
This is not stability.
It is delayed collapse.

Phase V: Distortion Accumulation

Without clean return, correction fails.
Errors compound across layers.
Misalignment increases between:
intention and outcome
decision and effect
Participants adapt locally to survive distortion:
compliance replaces authorship
narrative replaces clarity
justification replaces correction
The system maintains function while losing coherence.

Phase VI: Collapse and Misattribution

When accumulated distortion exceeds tolerance, failure occurs.
This may present as:
crisis
systemic breakdown
sudden loss of control
At collapse, the system attempts to assign responsibility.
Blame distributes downward or laterally:
operators are faulted
subsystems are replaced
individuals are removed
Yet the origin of direction remains unexamined.
Because authorship was obscured during operation, it cannot be recovered during failure.
The system cannot correctly locate responsibility.

Sovereignetics Conclusion

Authority without consequence is not achieved.
It is simulated.
Hierarchy does not remove responsibility.
It displaces it across time and structure.
The longer displacement is maintained, the greater the eventual return.
When consequence resolves, it does so proportionately to total participation, not perceived authorship.
The system fails not because hierarchy exists,
but because hierarchy was used to obscure the path between petition and return.
Consequence cannot be escaped.
Only delayed until it becomes unmanageable.

Applied Illustration: Relationship Betrayal

Consider a person who finds themselves repeatedly betrayed within an intimate relationship. The event is experienced as rupture: trust is broken, expectations collapse, and the emotional consequence is immediate and intense. The instinct is to locate cause externally, to identify the other as the source of harm and the explanation for what has occurred. This interpretation appears self-evident because the visible act of betrayal originates from outside the participant.

Within Sovereignetics, this surface reading is incomplete. The betrayal is not the beginning of consequence. It is its visible resolution. Prior to the event, a sequence of participation has already occurred. Signals were present. Inconsistencies were encountered. Boundaries were tested or softened. Decisions were made to remain, to defer, or to reinterpret what was observed. None of these moments carry the emotional intensity of the final rupture, but they form the structural pathway through which that rupture becomes possible.

A narrative that preserves authorship does not deny the betrayal. It does not excuse the other party or collapse the event into self-blame. It recognises the full chain of participation. It acknowledges that continuation occurred in the presence of contradiction, that action was not taken when tension first emerged, and that what followed is consistent with that pattern of engagement. This narrative does not resolve the discomfort of the situation. It retains it. It allows consequence to remain proportionate to participation.

Where this level of ownership cannot be sustained, the protective derivative intervenes.
Narrative shifts function. The event is reorganised into a form that reduces internal contradiction. The other becomes the primary cause. Language adjusts accordingly: manipulation is emphasised, inevitability is introduced, intention is softened or reframed. Statements such as "I had no choice," "I trusted too much," or "everything happens for a reason" begin to appear. These formulations stabilise the system. They reduce immediate tension and preserve a coherent sense of self.

What they do not do is resolve the structure that produced the outcome.
The narrative has changed, but the underlying participation remains unexamined. The Field does not respond to the explanation that has been constructed. It responds to the pattern that was enacted. Where that pattern is unchanged, consequence will reappear in different forms, often interpreted as coincidence, bad luck, or repeated misfortune.
The participant may then conclude that they "attract" certain types of people or that circumstances are beyond their control. This conclusion further reinforces the derivative narrative and deepens the displacement of authorship. The cycle persists not because the narrative is convincing, but because it has successfully stabilised the system without requiring structural correction.
The narrative resolved the feeling. It did not resolve the participation.

Ethics Under Derivative Authority

Ethical systems presuppose a coherent author capable of owning consequence. They assume that action is linked to reflection, that responsibility attaches to participation, and that proportional return can be integrated into behaviour.

Under derivative authority, these conditions do not hold.

When preservation governs participation, action is filtered before reflective authorship can engage. Ethical reasoning does not disappear. It is displaced. It operates downstream of preservation rather than upstream of participation.

Morality is not rejected.

It is repurposed.

Blame is externalised to reduce exposure. Consequence is deferred to maintain stability. Justification replaces integration. Narratives are constructed to protect orientation while preserving the appearance of coherence.

Ethical language remains intact.

Its function changes.

This is why individuals and institutions can articulate moral principles while repeatedly violating them. The failure is not ignorance of rules. It is structural misalignment. Ethics assumes ownership. Preservation restricts access to it.

Under these conditions, intent becomes easier to defend than outcome. Rules become easier to follow than proportion. Compliance becomes easier to maintain than coherence.

Ethical systems fracture not because they are false, but because they are executed by a structure organised around continuity rather than consequence.

As pressure increases, preservation tightens. Reflection narrows. Correction becomes threatening. Moral language persists, but it no longer regulates participation.

It protects identity.

This is why appeals to morality fail under stress. Systems experiencing threat do not evaluate coherence. They evaluate continuity. Ethical argument addresses authorship. The governing structure prioritises survival.

The result is performative ethics.

Values are declared. Principles are affirmed. Commitments are stated. Yet participation continues along preservation pathways. The distance between stated belief and enacted behaviour widens. Moral language becomes symbolic rather than operational.

The failure is mechanical.

Ethics requires authorship.

Derivative authority restricts access to it.

Until preservation is reordered beneath reflective participation, ethical systems will repeatedly fracture under pressure.

The problem is not insufficient principle.

It is structural inversion.

Responsibility cannot stabilise where exposure is experienced as annihilation.

Deferred Consequence

When the protective derivative governs, certain actions appear to escape consequence.

They do not.

They are deferred.

The system filters return, redirects cost, and distributes impact in ways that preserve short-term continuity. What would otherwise register as immediate proportion is postponed, displaced, or absorbed into background tension.

This produces the appearance of exemption.

An individual may act harmfully without immediate correction. An institution may exploit without visible collapse. A system may distort while remaining stable.

The absence of immediate return is misinterpreted as absence of consequence.

Nothing has disappeared.

What is not integrated is retained.

Deferred consequence stabilises as residual directionality—unprocessed tension embedded within participation. Behaviour continues. Surface stability is maintained. Distortion accumulates beneath it.

Delay does not reduce consequence.

It compresses it.

Because return is postponed, it accumulates without integration. When buffering capacity is exceeded, consequence re-enters abruptly. What appears disproportionate is accumulated proportion arriving without gradual release.

Collapse appears sudden.

It is not.

It is accumulated return becoming visible.

This explains why consequences surface long after originating action, why destructive patterns persist despite awareness, and why systems that appear stable for extended periods destabilise rapidly.

Delay is mistaken for escape.

It is storage.

The derivative does not eliminate consequence. It manages timing.

At scale, deferred consequence appears as generational burden, systemic fragility, and saturation events. Cost may be displaced across individuals, groups, or timeframes, but redistribution does not remove return.

The geometry of consequence remains intact.

Perceived exemption depends on distance. When the link between action and return is stretched, authorship appears diluted. Responsibility appears negotiable.

This is an illusion.

The Field does not register delay as absolution.

It registers projection.

When buffering fails, stored distortion re-enters participation. The intensity of return reflects the duration and density of deferral.

What appears as collapse is consequence regaining access.

This process is not punitive.

It is structural.

Preservation without integration accumulates instability. The more aggressively consequence is deferred, the more constrained the system becomes.

What is postponed does not dissipate.

It concentrates.

Responsibility without exemption requires recognising this mechanism. The absence of immediate feedback is not evidence of innocence. Delay is not evidence of safety.

What appears as exemption is consequence awaiting return.

Why Insight Fails

Understanding is not integration.

Under derivative authority, insight occurs at the level of narrative. Behaviour can be analysed, patterns identified, history reconstructed, and consequences articulated. This creates the impression that change is imminent.

It is not.

Insight operates in language.

Participation is governed by orientation.

A system may fully recognise its distortion and continue enacting it because the structure that initiates behaviour has not shifted. Awareness does not reach the level at which participation is generated. The narrative layer observes.

The system continues.

Morality fails for the same reason.

Ethical systems address reflection. They assume a participant capable of responding to principle. Values are prescribed, duties defined, alignment expected to follow. Under derivative authority, behaviour is not governed by principle. It is governed by preservation.

When moral demand requires reorientation that feels destabilising, it is interpreted as threat. Defensive logic overrides ethical reasoning. Agreement is possible.

Change is not.

The failure is not hypocrisy.

It is structural mismatch.

Reframing intensifies this effect. Interpretation can be altered, narratives reconstructed, meaning reassigned. Distress may reduce. Coherence may appear restored.

Orientation remains unchanged.

Reframing modifies description.

It does not modify direction.

This is why cycles persist despite awareness. Insight accumulates without altering participation. Knowledge reinforces identity rather than shifting behaviour.

The same pattern appears at scale. Institutions articulate values, publish frameworks, conduct audits, and refine language. Insight becomes formalised.

Operational structure remains intact.

Narrative evolves.

Direction does not.

Because preservation governs, conceptual adjustment is tolerated as long as continuity is not threatened. Interpretation can change. Exposure is resisted.

Reorientation requires something else.

Consequence must be encountered without collapse. Participation must remain intact through destabilisation. Only then does the premise that sustains defensive structure begin to weaken.

Until then, insight stabilises distortion. Morality justifies it. Reframing sustains it.

The Field does not respond to interpretation.

It responds to participation.

Insight without reorientation accumulates tension. Moral agreement without structural shift compounds contradiction. Reframing without exposure delays correction.

This is why lasting change often follows disruption rather than understanding.

Comprehension is tolerated.

Unfiltered consequence is not.

Where consequence is not fully received, integration does not occur.

Without integration, direction does not change.

The Derivative Cannot Be Removed

The protective derivative cannot be removed.

It cannot be eliminated, dissolved, transcended, or integrated out of existence. This is not a therapeutic claim.

It is structural.

The derivative exists because participation occurs through finite vessels. Embodied systems are vulnerable, exhaustible, and subject to termination. Where participation is exposed to potential collapse, a mechanism for preservation emerges.

The derivative is that mechanism.

It is not an anomaly.

It is a consequence of finitude.

Its removal would require conditions that do not exist within participation: certainty of continuity while embodied, consequence without vulnerability, exposure without risk. These conditions are unavailable within constrained systems.

Any framework that promises complete removal of defensive structure assumes what cannot be provided.

The derivative is not an error in the system.

It is the system's response to limitation.

What can change is not its existence. It is its position.

When the derivative functions as an instrument, it preserves stability without displacing authorship. It responds to genuine threat while allowing reflection to recalibrate participation.

When it becomes dominant, preservation replaces reflection. Continuity overrides proportion. Participation narrows.

The distinction is not between presence and absence.

It is between use and governance.

Attempts to eliminate the derivative misidentify its role. Suppression intensifies it. Denial reinforces it. Narratives of transcendence often reproduce the same preservation logic under different language.

The task is not removal. It is reordering.

The derivative must remain available for conditions of real risk, but it must not organise participation in their absence. Preservation must operate beneath reflection, not above it.

As long as participation occurs through vulnerable structures, protection will arise.

The question is not whether the derivative exists. The question is whether it governs.

Responsibility becomes accessible only when preservation serves authorship rather than replacing it.

Epistemic Speed and the Derivative

Not all explanations stabilise participation at the same rate. Some require reflection. Others resolve tension immediately.

When continuity risk is present and consequence is difficult to track, the system becomes highly receptive to rapid stabilisation. Explanations that reduce exposure to consequence are adopted more quickly than those that require engagement with it.

This produces epistemic speed asymmetry.

Conceptual closure can occur faster than reflective authorship. An explanation may feel complete before the system has encountered the consequence it describes. Tension resolves at the level of interpretation, not participation.

Frameworks that terminate responsibility are especially efficient in this environment. Explanations that appeal to inevitability, abstraction, distributed agency, or external causation reduce the perceived burden of authorship. They provide immediate orientation without requiring recalibration of participation.

The effect is stabilisation without exposure.

Relief is often mistaken for accuracy. When tension dissolves quickly, the explanation that produced that relief appears valid. The system does not need to test it against consequence because the pressure that would drive that testing has already been reduced.

This is why rapid diffusion is not evidence of coherence.

Explanations that spread quickly tend to align with existing defensive orientation. They stabilise interpretation in a way that preserves continuity while minimising confrontation with consequence. Repetition and adoption reinforce the appearance of validity, but what is spreading is resonance, not verification.

Explanations that require reflection move more slowly. They delay closure. They reintroduce authorship where it has been displaced. They expose participation to consequence rather than resolving it prematurely. This slows adoption because it increases structural demand.

The difference is not intellectual complexity.

It is exposure.

Fast explanations close the system before responsibility forms. Slower explanations leave the system open long enough for consequence to be encountered.

This is why the point at which an explanation feels complete is often the point at which authorship has been displaced.

Completion is not proof of truth.

It is often the boundary at which responsibility has been avoided.

The Epistemic Derivative

The protective derivative does not operate only in behaviour. It operates in interpretation.

When continuity risk governs participation, it must also shape how reality is explained. If interpretation remained unfiltered, reflection could restore authorship. To prevent this, preservation extends into epistemology.

What is seen, selected, and concluded is shaped before reflection fully engages.

At the epistemic level, three functions emerge.

Destabilising information is abstracted.
Authorship is replaced with inevitability.
Responsibility is relocated into systems, laws, or universals.

This is how philosophical insulation stabilises.

When exposure to consequence feels threatening, explanations that dissolve authorship become attractive. Determinism renders events unavoidable. Simulation reframes action as externally generated. Structural explanations relocate cause into impersonal systems. Historical narratives convert participation into momentum. Each reduces exposure.

The issue is not descriptive accuracy. Many such frameworks capture real patterns.

The issue is termination.

They explain structure while displacing authorship within it.

The system can understand reality without re-entering responsibility.

Under these conditions, epistemic speed replaces reflective depth. Explanations that resolve tension quickly feel convincing because they align with defensive urgency. Frameworks that conclude with inevitability mirror the system's internal orientation: that events could not have unfolded otherwise.

Universals reinforce this process. If gravity governs, biology determines, economics compels, or history necessitates, participation is positioned downstream of structure.

Responsibility diffuses through description.

The derivative does not create these explanations. It selects them. It stabilises within interpretations that preserve continuity and resists those that restore authorship.

This is why certain frameworks persist despite contradiction. Their function is not only explanatory.

It is protective.

They reduce burden while allowing participation to continue without full exposure to consequence.

As this stabilises, inquiry shifts orientation. Investigation seeks confirmation rather than coherence. Evidence becomes shield. Consensus becomes validation. Explanation becomes refuge.

The mirror is mistaken for authority because authority removes the need for ownership.

This is not conscious manipulation. It is structural bias.

At scale, it becomes institutionalised. Academic paradigms resist reorientation. Political narratives harden into inevitability. Technological trajectories are framed as unstoppable. Collective continuity risk produces collective abstraction.

Authorship dissolves before it becomes actionable.

Sovereignetics interrupts this. It does not deny structure, constraint, or pattern. It denies that structure absorbs authorship.

When explanation removes responsibility, it has shifted from description to insulation.

The derivative stabilises within that shift.

Authorship returns only when explanation no longer functions as refuge.

Until then, knowledge itself becomes delegated authority.

Participation continues. Ownership does not.

Why Derivative Systems Persist

Derivative systems do not collapse immediately. They persist, and this persistence creates the illusion that responsibility has been successfully displaced. It has not. Their short-term stability is maintained through mechanisms that alter the visibility and pacing of consequence without removing it.

The first of these is delay of return. When consequence is displaced, it does not disappear; it lengthens its path. Buffers, layers, and intermediaries increase the time between action and return, weakening perceived causation. Actions begin to appear disconnected from outcomes, and decisions appear insulated from consequence. The system continues to function while misalignment accumulates beneath the surface.

The second mechanism is distribution of impact. Consequence is not removed; it is spread. Multiple participants absorb fragments of return, preventing any single locus from experiencing the full proportion. This diffusion prevents immediate collapse. Each participant adapts locally, adjusting behaviour, compensating for distortion, and absorbing inconsistency. The system survives, but only through distributed strain.

The third mechanism is narrative stabilisation. Where consequence becomes unclear, explanation replaces visibility. Narratives form to justify outcomes, with failure attributed to individuals, success attributed to structure, and causation reframed to protect origin. These narratives do not resolve distortion; they prevent recognition of it.

The system appears coherent because its explanation is coherent, not because its structure is.

Together, these mechanisms create temporary viability. Delay masks causation, distribution reduces intensity, and narrative suppresses detection. The system continues without immediate correction, but this stability is conditional because consequence remains cumulative.

What is delayed compounds, what is distributed converges, and what is obscured reappears under saturation.

Collapse does not occur when error is introduced. It occurs when accumulated distortion exceeds the system's capacity to absorb it. At that point, delay collapses into immediacy, distribution collapses into concentration, and narrative collapses into irrelevance. Return becomes unavoidable and disproportionate to any single visible cause.

This is why derivative systems fail suddenly. They do not deteriorate linearly; they maintain function while coherence degrades. When collapse occurs, it appears abrupt, but structurally it is the completion of accumulated consequence.

Derivative systems persist because consequence can be delayed. They fail because consequence cannot be removed.

Part VI: Responsibility Beyond the Individual

Sovereign Responsibility

Sovereignty is often mistaken for isolation, my will, my choice, my life.

In Sovereignetics, sovereignty is not separation from impact.

It is authorship inseparable from consequence.

Ache is sovereign at its centre, but its petitions do not remain contained within the vessel that projects them. They enter a shared Field where reflection interacts, propagates, and returns. Responsibility therefore does not terminate at intention.

It persists until consequence stabilises.

Most ethical frameworks impose boundaries on responsibility. It may end at intention, awareness, legality, or direct causation. Each boundary provides insulation.

Sovereignetics rejects them.

The Field is shared. Reflection is relational. Consequence extends beyond the originating act. An action is not complete when it is chosen.

It is complete when its return resolves.

Every petition alters Field geometry. That alteration shifts probability, conditions local coherence, and influences downstream participation. Distortion does not remain contained. It propagates through relational structure, normalising itself, lowering resistance, and redistributing corrective load.

This is not exaggeration.

It is propagation.

Collective structures do not dilute authorship.

They scale it.

Institutions, cultures, and systems are configurations of participation. They do not replace locality; they extend it. Appeals to abstraction, policy, culture, system, do not remove authorship.

The Field does not recognise abstraction.

It recognises participation.

Shared action produces shared saturation, not dissolved responsibility.

Responsibility is not guilt.

It is attachment.

Consequence follows participation regardless of intention, awareness, or acknowledgement. Indirect effects do not disappear. Ignorance does not interrupt return. The Field reflects what is enacted, not what is recognised.

Responsibility also extends to sustained benefit. To profit from distortion, to rest upon extracted coherence, or to maintain advantage without repair is to continue participation in that distortion.

Benefit is not passive.

It is ongoing projection.

Non-action is not neutral. Silence alters circulation. Withdrawal stabilises what remains unchallenged. To witness distortion without response, to benefit without correction, or to recognise incoherence without ownership is to participate in its continuation.

Permission carries consequence.

Because return can be delayed, responsibility can outlast awareness. It persists through denial and reappears under new conditions. This is not moral inheritance.

It is directional continuity.

Repair is not moral penance.

It is structural completion.

To repair is to restore circulation where extraction occurred, to integrate where distortion was projected, and to close what remains open. Refusal to repair narrows participation. Engagement restores it.

Responsibility intensifies with clarity. As perception sharpens, authorship becomes explicit. Power without corresponding responsibility destabilises rapidly because participation exceeds ownership.

Insight does not reduce obligation.

It amplifies it.

Sovereign responsibility extends wherever participation reaches, through relationship, structure, and time. It does not dissolve at the boundary of the individual.

It persists until consequence stabilises.

To be sovereign is not to escape responsibility.

It is to recognise that participation and consequence remain coupled beyond intention.

The Bomb Thought Experiment

Responsibility is often mislocated at outcome rather than authorship. Moral reasoning, legal language, and social discourse frequently tie responsibility to visible harm, measurable damage, or proven intent.

The bomb thought experiment removes this assumption.

An individual places a bomb. It is real, functional, and armed. It will detonate, though the timing is uncertain. It may explode in minutes, days, or years. No one else knows it exists.

The individual leaves.

The immediate question is when responsibility begins. At detonation? At harm? At fear? At proof of intent?

Each assumes responsibility attaches to events. It does not.

Responsibility begins when the device is armed.

At that moment, consequence has been introduced into the Field. A trajectory has been authored. Direction now exists independent of when it becomes visible.

Delay affects visibility. It does not affect origin.

If responsibility depended on outcome, postponement would dilute authorship. A bomb set to detonate years later would carry less responsibility than one detonating immediately.

This is incoherent.

Time alters experience. It does not alter authorship.

The one who armed the device authored the trajectory regardless of when consequence appears.

Responsibility attaches to direction.

Not to explosion.

This is destabilising because it removes temporal insulation. Harm does not need to materialise for responsibility to exist. Risk, once introduced, is already participation.

The structure generalises.

Environmental destruction, institutional negligence, economic exploitation, technological deployment, each can function as delayed consequence. Harm may be distributed, postponed, or obscured.

Direction has already been set.

Responsibility does not begin when consequence becomes visible.

It begins when consequence becomes inevitable through participation.

Time does not erase direction.

It separates projection from recognition.

Integrity Under Constraint

Sovereignetics does not divide beings into righteous and wicked.

It distinguishes only coherence and incoherence in participation.

Constraint narrows participation. It compresses available action, limits expression, and reduces the range through which consequence can be enacted. Oppression is one form of this compression. Participation continues within whatever geometry remains.

Constraint does not degrade Ache.

It concentrates it.

Under pressure, participation may fragment. Identity may collapse into survival. Distortion may be internalised. Coherence may be mimicked for protection.

But another trajectory remains possible.

Participation can maintain alignment within constraint.

When this occurs, coherence stabilises internally even while external conditions remain hostile. This is not because suffering is virtuous. It is because participation continues without surrendering orientation.

Ache does not adopt imposed distortion.

It remains aligned with its own projection.

This produces structural stability. Identity becomes less dependent on external validation, less reliant on domination, and less fragmented by fear. The vessel remains constrained, but the internal architecture strengthens.

Constraint tests alignment.

Power tests dependence.

Where participation expands through asymmetry, access increases. Authority, resources, and insulation from immediate consequence widen available action. Under these conditions, fragmentation is less likely.

Dependency is more likely.

Identity may stabilise through control. Validation may be sourced through compliance. Coherence may be replaced by dominance. What appears as strength is often contingent on maintaining the asymmetry that supports it.

This is not coherence.

It is dependence.

Remove the asymmetry and the structure destabilises. Participation that relies on extraction cannot sustain itself. It requires continuous input from others to maintain stability.

This is parasitic participation.

It appears strong while conditions hold.

It collapses when they change.

The divergence between constraint and power is not moral.

It is structural.

Constraint compresses participation. Power expands it. In both cases, outcome depends on orientation.

Constraint does not produce coherence automatically. Power does not produce distortion automatically.

Alignment under pressure stabilises participation.

Extraction under freedom destabilises it.

The Field does not judge these trajectories.

It reflects them.

Participation aligned with coherence expands capacity. Participation built on distortion narrows until correction becomes unavoidable. No external authority intervenes.

Geometry is sufficient.

Oppression does not sanctify.

Power does not corrupt by necessity.

What matters is orientation within condition.

Where alignment persists under constraint, capacity strengthens. Where participation feeds on distortion, it becomes dependent and fragile.

The divergence is not ethical.

It is structural.

Ethics Under Constraint

Ethics do not exist independently of condition.

They stabilise within it.

When participation occurs under sufficient stability, coherence can be maintained. Behaviour aligns with stated values because the cost of alignment remains tolerable. Under these conditions, ethics appear stable.

They are not.

They are supported.

When constraint intensifies, participation reorganises. Scarcity, threat, and instability compress available geometry. Consequence becomes immediate. Survival becomes dominant. Under these conditions, previously stable behaviour can shift rapidly.

This is not hidden immorality.

It is structural pressure.

Ache does not suspend participation under constraint.

It redirects it.

A participant may value fairness and still exploit when survival appears threatened. Cooperation may collapse into competition. Law may be abandoned when structure fails. These are not contradictions in identity.

They are reorientations under pressure.

Constraint does not remove ethics.

It tests their structural viability.

When consequence becomes acute and buffering disappears, participation narrows toward preservation. Coherence requires capacity. When that capacity is exceeded, alignment destabilises.

This is not justification.

It is mechanism.

Ethics cannot be sustained by intention alone. They require conditions in which coherence remains possible. When systems fail, those conditions disappear. Participation reorganises around what can be maintained within the available geometry.

Survival replaces alignment.

This reveals a structural dependency. Ethical stability is not solely an individual property. It is supported by environment, structure, and distribution of consequence.

When these fail, behaviour follows.

The Field does not evaluate the shift.

It reflects it.

Participation under constraint still generates consequence. Reorientation does not remove return. It alters trajectory.

Ethics under constraint therefore expose the limits of unsupported coherence. Where conditions cannot sustain alignment, participation narrows until correction becomes unavoidable.

The issue is not whether ethics exist.

It is whether they can be maintained under pressure.

Distributed Action

Responsibility is often misinterpreted as proportional to scale. Small actions are treated as negligible. Individual contributions within large systems appear diluted. Harm that emerges slowly or collectively is therefore perceived as belonging to no one.

This confuses visibility with authorship.

Distributed action does not dissolve responsibility.

It multiplies participation.

Large-scale outcomes rarely arise from a single act. They emerge through accumulation. Policies stabilise through repeated approvals. Environments degrade through continued extraction. Cultural distortions persist through countless permissions. Each contribution appears minor in isolation.

In aggregate, they form trajectory.

The absence of immediate feedback does not negate participation. When consequence is delayed, return is extended, not removed. Distributed harm often moves through complex structure before localising. By the time it appears, it seems detached from origin.

It is not.

Detachment is interpretive.

Each participant authors within their available geometry.

No individual may control the entire structure, yet each act that sustains it reinforces its direction. Participation compounds. Outcome reflects alignment density, not individual dominance.

The appeal of distribution is diffusion. When many contribute, exposure feels reduced. Responsibility appears fractional. "I was only one part" becomes insulation.

Proportion does not divide.

Participation either reinforces trajectory or interrupts it.

Delay intensifies this illusion. When return emerges far from the point of action, the connection appears abstract. Time creates distance in perception.

It does not alter structure.

Distributed action combined with delayed consequence produces cumulative distortion. Harm stabilises without immediate correction. Participation normalises because no single act appears decisive. When return finally localises, collapse appears sudden.

It was accumulated.

This applies equally to coherence. Alignment compounds. Repair stabilises through repetition. Small acts of coherence reinforce structure just as small acts of distortion degrade it.

The Field does not privilege scale.

It tracks density.

Aligned participation strengthens structure. Misaligned participation destabilises it. Outcome reflects accumulation, not visibility.

Responsibility intensifies with awareness. Once trajectory is recognised, continued participation carries amplified authorship. Knowledge removes the illusion of insignificance.

No participant authors the entire system.

Every participant contributes to it.

Distributed action does not erase responsibility.

It distributes participation.

Return remains coupled to what is reinforced.

When consequence localises, it reveals what participation has been building.

Fragmented Trust

Trust is often treated as a virtue, something to be extended, preserved, or repaired as a moral obligation.

When framed this way, it detaches from structure.

It becomes identity.

Trust is not a belief.

It is a consequence-tracking mechanism.

It emerges from observed consistency, alignment between participation and return, and exposure to proportion. It is the expectation that participation will remain coherent.

Fragmentation occurs when trust is abstracted into principle rather than maintained as calibration. Statements such as "trust others," "trust is necessary," or "distrust is a flaw" convert trust into a moral position.

Once moralised, it resists correction.

Questioning it becomes identity threat rather than recalibration.

When trust detaches from consequence, it overrides proportion. Warning signals are dismissed. Inconsistency is rationalised. Boundary enforcement is reframed as betrayal. Trust ceases to track participation.

It begins protecting itself.

Failure follows.

Exploitation occurs. Distortion accumulates. Systems destabilise. Yet response rarely addresses the structure. It swings between naïve re-extension and total withdrawal.

Both repeat the error.

The failure was not trust.

It was exemption.

Trust treated as a universal becomes insulated from consequence. It demands continuation without recalibration. It functions as abstraction rather than mechanism.

In this state, it becomes parasitic.

It protects comfort at the expense of coherence.

Trust functions only when it remains tethered to participation and return. It must update continuously. It must respond to proportion. Without recalibration, it no longer tracks the Field.

It distorts it.

This is not unique to trust. Any construct detached from consequence becomes unstable. Trust simply exposes the mechanism clearly.

When trust stops tracking participation, it stops tracking reality.

The Purpose of Life in Sovereignetics

One of the most enduring questions human beings ask is simple:

What is the purpose of life?

Religious traditions answer by appealing to divine intention, commandments, and ultimate judgment. Spiritual traditions often speak of awakening, ascension, or reunion with a higher consciousness. Secular frameworks substitute progress, happiness, success, or self-actualisation. Despite their differences, these explanations share a common assumption: that life exists in order to accomplish some pre-assigned objective.

Sovereignetics rejects that assumption.
Existence is not a lesson. It is not a test. It is not a punishment, and it is not preparation for another realm. There is no external authority assigning a mission to each being, and no hidden criteria against which participation is scored.

The purpose of life, in Sovereignetics, is participation.

This is not a slogan. It is a structural conclusion.

Ache is the irreducible will-to-exist and will-to-become. It does not enter reality to fulfil a cosmic assignment. It participates because participation is what Ache does. To exist is to press into the Field and receive consequence in return. There is no position outside this process.

Participation is not optional; it is the condition of existence itself.

This answer may initially seem unsatisfying to those who expect purpose to be grand or transcendent. But its simplicity is precisely its strength. The meaning of life is not hidden behind metaphysical abstractions. It is already occurring; You are here to participate.
Every thought, action, omission, commitment, and withdrawal alters the Field and returns consequence. Whether participation is coherent or fragmented, conscious or automatic, consequence follows. The central question is not whether you are participating. The central question is how.

Growth, learning, and evolution are not the purpose of life. They are possible outcomes of coherent participation. When a being engages reality with increasing clarity and alignment, consequence tends to become more proportionate and intelligible. Stability increases. Distortion diminishes. Insight deepens. This is commonly experienced as growth. But growth is not guaranteed, and it is not demanded by the universe. It is what often occurs when participation becomes more coherent.

Meaning arises in the same way. Meaning is not embedded in the universe as a pre-existing property waiting to be discovered. It emerges when a participant reflects upon experience and integrates the patterns revealed by consequence.
Meaning is real, but it is derivative. It arises from participation; it does not precede it.
This conclusion also dissolves nihilism.

Nihilism assumes that if there is no externally assigned purpose, then nothing matters.
Sovereignetics reaches the opposite conclusion.
Things matter precisely because consequence is real.
Every mode of participation alters what returns.
Reality does not need to provide a cosmic mission for existence to be significant. Significance is built into the structure of participation itself.

Nor are we "just patterns" in any dismissive sense.
We are loci of Ache through which reality becomes experienced and shaped.
We are bounded centres of participation whose choices influence the quality and direction of consequence. That is not trivial. It is the basis of authorship.

From this perspective, self-improvement is not about fulfilling a universal ideal. It is about participating more coherently. To become more truthful, disciplined, compassionate, or courageous is to reduce distortion and improve the fidelity with which consequence is encountered and integrated.

The practical implication is straightforward:

Participate as coherently as possible.
This is not a commandment.
It is a structural recommendation.
Coherent participation tends to produce clearer, more stable, and more proportionate returns. Incoherent participation tends to produce friction, confusion, and collapse.

Sovereignetics therefore offers a direct answer to one of humanity's oldest questions.

The purpose of life is participation.

Growth is what often follows coherent participation.
Meaning is what emerges when experience is integrated.
And consequence is what ensures that how you participate always matters.

Closing: The Persistence of Consequence

Consequence does not disappear.

It reorganises.

Responsibility is not a psychological state and not a moral burden that can be accepted or rejected. It is not something one takes on or lets go. It is the structural continuity between participation and return. Once participation occurs, geometry is altered.

That alteration does not revert.

What is authored persists as orientation.

Forgetting does not erase it. Reframing does not dissolve it. Justification does not neutralise it. Silence does not suspend it. Responsibility displaced into systems does not absorb it.

Consequence is not negotiated through narrative.

It is stabilised through structure.

The Field does not require acknowledgment to reflect proportion. It does not wait for confession, agreement, or recognition. Reflection has already occurred at the level of participation.

Every act leaves direction.

That direction may stabilise as coherence, accumulate as distortion, remain as tension, or compound until saturation.

It does not vanish.

There is no mechanism within participation that deletes orientation without integration.

This is what destabilises.

Most ethical systems preserve the possibility of erasure. They offer pardon, redemption, closure, reform, or transcendence as endpoints.

Sovereignetics does not deny repair.

It denies deletion.

Repair integrates. It does not erase.

What is integrated becomes capacity. What is avoided becomes constraint. What is deferred narrows future participation. These are not moral outcomes.

They are structural consequences.

Even death does not cancel authorship. The end of a vessel terminates a configuration of experience.

It does not rewrite participation.

Pattern remains pattern. Orientation remains orientation. Reset clears arrangement.

It does not annul direction.

There is no archive where consequence disappears.

There is only transformation.

Nothing within a reflective Field evaporates. It circulates, integrates, persists, or collapses into new form. The illusion of exemption depends on believing that consequence is temporary, localised, or erasable.

It is none of these.

Responsibility without exemption means that participation continues in its effects even when awareness does not. Denial does not interrupt return. Distance does not dissolve authorship.

The Field continues.

There is no outside of consequence.

No position from which participation can be undone. No abstraction into which return dissolves. No delay that converts authorship into innocence.

The only variable is direction.

Whether orientation integrates.

Or accumulates.

Persistence alone is not the full structure.

What remains continues.

Orientation accumulates as momentum.

Momentum becomes trajectory.

Trajectory stabilises participation.

Coherence expands capacity.

Distortion narrows possibility.

Unresolved tension directs what participation becomes.

The Field does not only reflect what has occurred.

It conditions what becomes possible next.

Responsibility establishes that consequence cannot be escaped.

The next question is what consequence becomes.

Volume III begins there.

The question is not whether responsibility applies.

The question is what you are stabilising.

Leave a Review

If this work challenged your assumptions, clarified something important, or shifted the way you understand reality, consider leaving a brief review on Amazon.

Your review does not need to be technical or comprehensive. Even a few sentences about what stood out to you, what you found valuable, or how the work affected your thinking can help other readers decide whether to engage with Sovereignetics.

This system is not built on agreement. It is built on engagement.

Thank you for taking the time to read and respond.

Continue the Sovereignetics Series

For additional volumes, collected editions, and reference works, search:

J. Anthony De Angelis – Sovereignetics